BIOLOGY FOR
THE BEGINNER

BIOLOGY FOR THE BEGINNER

In association with Atif Elnaggar, PhD,
professor of biology

NICHOLAS J. ORME M.D.

To order additional copies of this book, contact:
Xlibris
844-714-8691
www.Xlibris.com
Orders@Xlibris.com
840982

Contents

Introduction

In this book we will strive to meet nine goals:

First we will define life. What characteristics must an organism possess to be alive? How did life start on Earth? Why Earth? Why don't the other seven planets support life? What had to happen for unicellular life to appear?

Then we will look at the smallest unit of life: the cell. What does it look like? Where does it come from? What is it made of, and how does it work? How does it differentiate into an adult? How does it produce special substances? How does it reproduce? And why does it eventually die? How did single celled organisms evolve into multi cellular life?

How did living things evolve into the six million life forms that live on Earth today? What caused living things to leave the oceans and inhabit land? When did humans appear, and why were they able to accomplish so much? How do biologists classify living organisms?

Then we will discuss anatomy and physiology. What organ systems do higher life forms possess, and how do they work? We will discuss genetics. How do living things inherit characteristics from the parents, and how do they adapt and evolve?

We will discuss some of the medical conditions that affect higher life forms and what they need from their environment. We will discuss the unique gifts that living organisms possess. Finally, we discuss factors that endanger life and what we need to be careful about in the future.

I hope you find this to be as enjoyable as I did.

BIOLOGY FOR THE BEGINNER

By Nicholas J. Orme, M.D.

In association with Atif Elnaggar,
PhD, Professor of Biology

Definition of Life

Let us begin by defining life. To be alive you must:

> be able to reproduce
> consume nutrition
> grow
> respond to stimulus
> adapt
> evolve
> be made of molecules that contain carbon

Up until the Middle Ages people believed in "spontaneous generation," the idea that life could arise from inanimate objects or be created artificially. Grains of wheat could spontaneously turn into mice or meat could become flies. Carlo Colloid thought that a wooden doll could turn into a boy, and in 1883 he wrote *Pinocchio*.

But in 1666 Francisco Redi told the world that "life could only arise from preexisting life" and proved it through a series of brilliant experiments. He took two flasks with broth and exposed them to the air. He left one flask open and covered the other with cheesecloth. Air was able to enter both and all other conditions were identical. The flask that was open developed mold, whereas the one with the cheesecloth did not. So life was only possible in the flask where a preexisting organism was able to access the broth. Without this, the broth could not produce life.

Let us move on to Stanley Miller's theories about Earth's early atmosphere. Stanley Miller was a professor of chemistry at the University of Wisconsin. In 1953 he suggested that Earth's early atmosphere was composed of ammonia, water vapor, nitrogen, hydrogen, hydrogen sulfide, methane, and carbon dioxide—similar to modern-day Venus.

Earth is in what astronomers refer to as the Goldilocks region relative to its sun: not too cold and not too hot but just right,. just right for water to exist as a liquid and not exclusively as ice or as steam. Billions of years ago the atmosphere cooled to the point of allowing water to exist as a liquid. This precipitated two events: it created the oceans and triggered millions of years of thunderstorms. Dr. Miller theorized that lightning converted the primordial atmosphere into twenty-one amino acids. He theorized that the amino acids mixed with the oceans, which were hot, and somehow assembled into early life.

One of the earliest life forms to appear on Earth were the Prokaryotes; single celled organisms that reproduced by binary fission and whose DNA was distributed throughout the cell. They gave rise to the Eukaryotes whose DNA was confined to a nucleus.

Some eukaryotes exchange genetic matter through conjugation. They connect with one another through a probosus (a tube). Some single celled organisms also form giant colonies called volvox. This might have been the earliest form of sexual reproduction and multicellular life.

Once the oceans formed, they further regulated the temperature. If it was too hot, they would evaporate and cool the environment. If it was too cold, it would rain and snow. This, along with the four seasons, regulated the temperature of the Earth.

Dr. Miller built the apparatus that appears on the preceding page. It recreated Earth's early atmosphere and passed electric sparks through it. Then he cooled it until it formed a liquid. When he analyzed the liquid, he found that it contained twenty-one amino acids.

Stanly Miller professor of chemistry

All   **Images**   Videos   Any License

Color
| All |

Size
| All | S | M | L |

Type
| All | Photo | Graphics | GIF | Face | Portrait | Non Portrait | Clipart | Line Draw |

Showing results for *stanley Miller professor of chemistry*
Search instead for Stanly Miller professor of chemistry

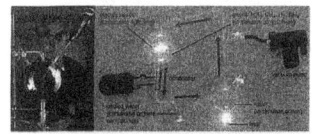

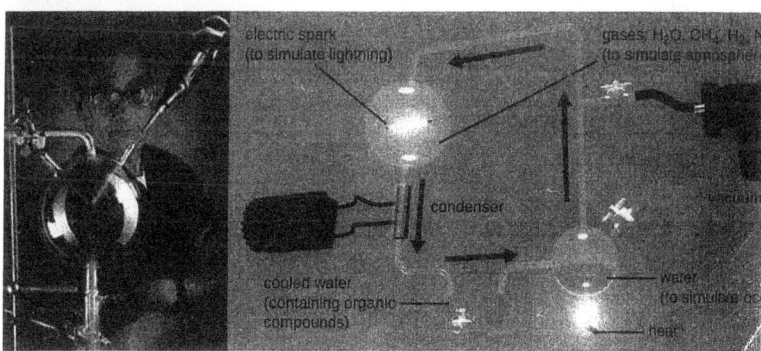

The Cell

The cell is the smallest unit of life. Some cells live independently, but higher life forms consist of tissues—a collection of similar cells that perform a common function. Single-celled life forms include:

amoeba
protozoa
fungi
algae
molds
bacteria

When the oceans were filled with blue-green algae, they produced oxygen through photosynthesis. That created the oxygen and nitrogen atmosphere that we breath today. This made multicellular life possible. Organisms that used oxygen had more energy available than those that did not, and they used it to create multicellular life.

G. Benaglia inc.

Francisco Reid.

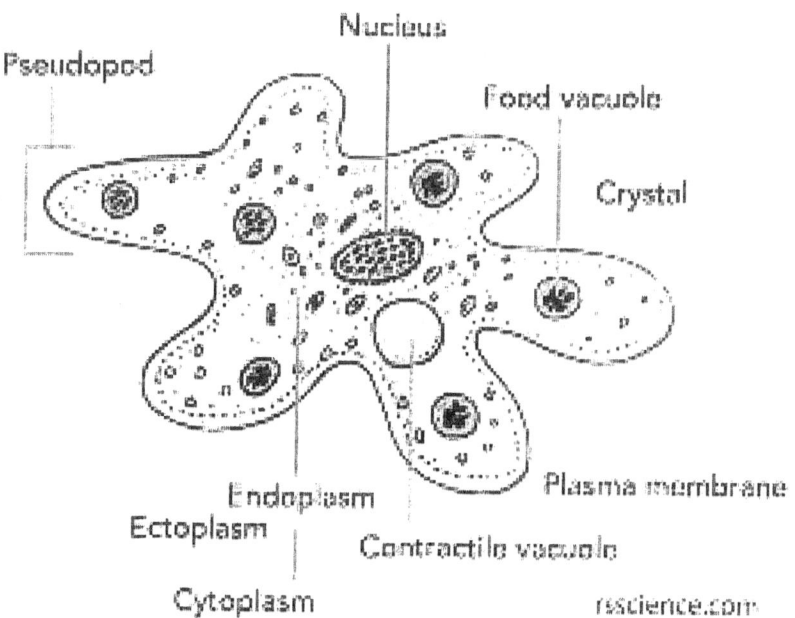

Amoeba.

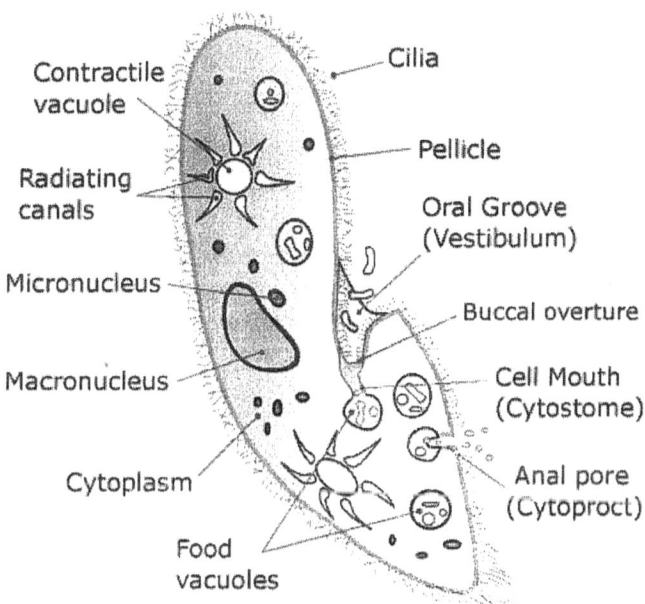

Contractile vacuole

Cilia

Radiating canals

Pellicle

Oral Groove (Vestibulum)

Micronucleus

Buccal overture

Macronucleus

Cell Mouth (Cytostome)

Cytoplasm

Anal pore (Cytoproct)

Food vacuoles

Protozoa.

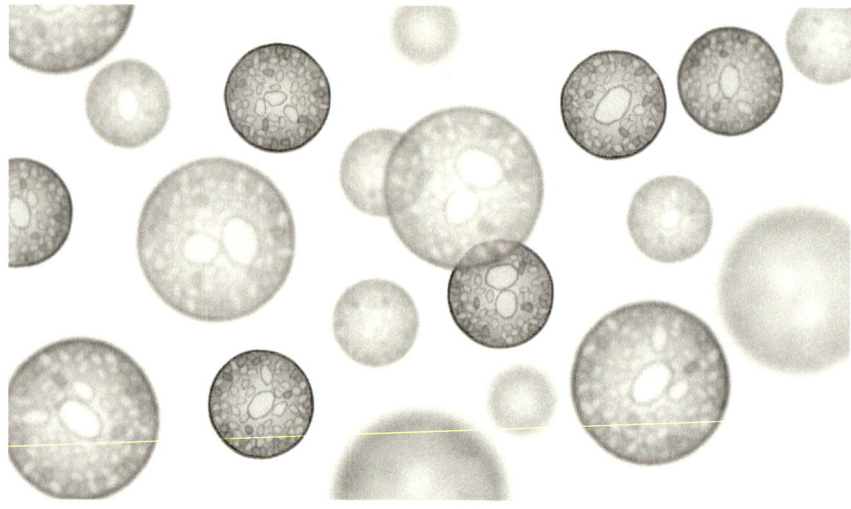

Algae.

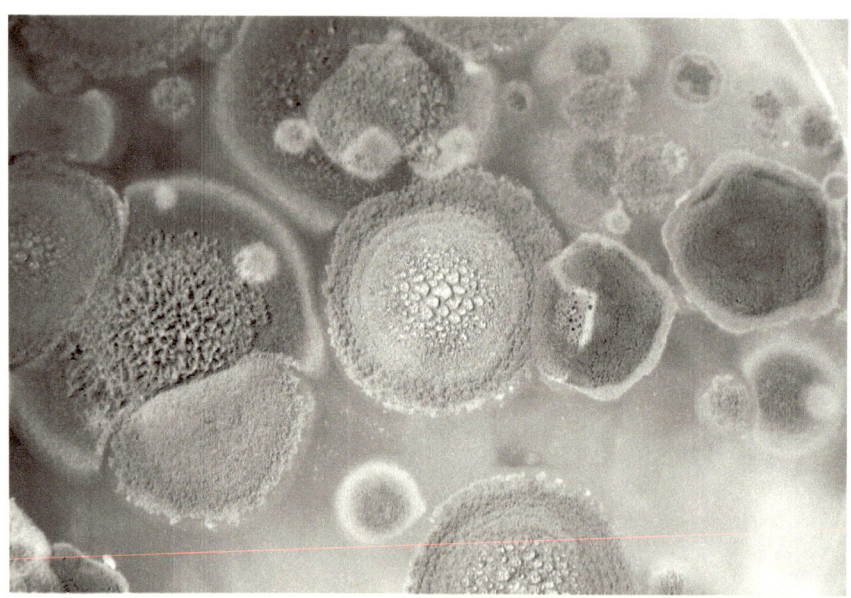

Mold.

Fungi.

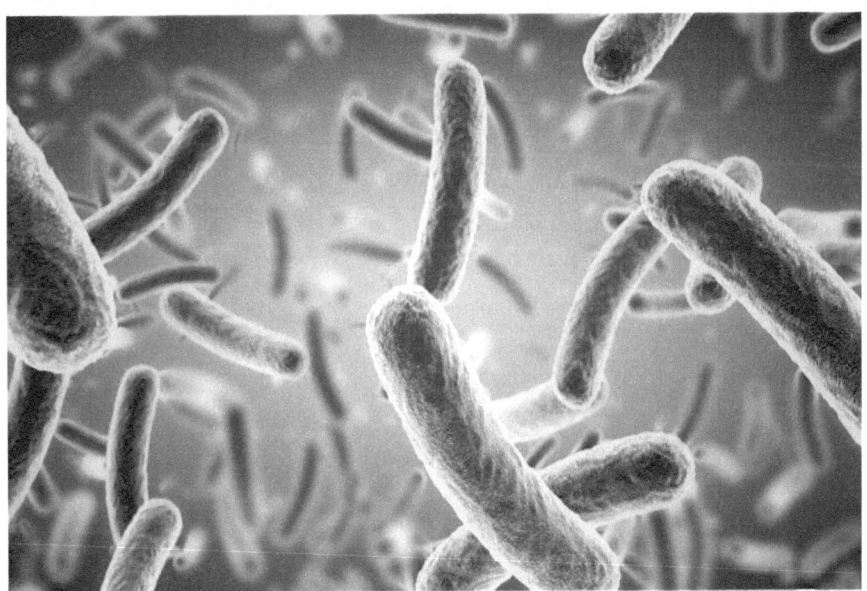

Bacteria (Bacillus).

Binary Fission

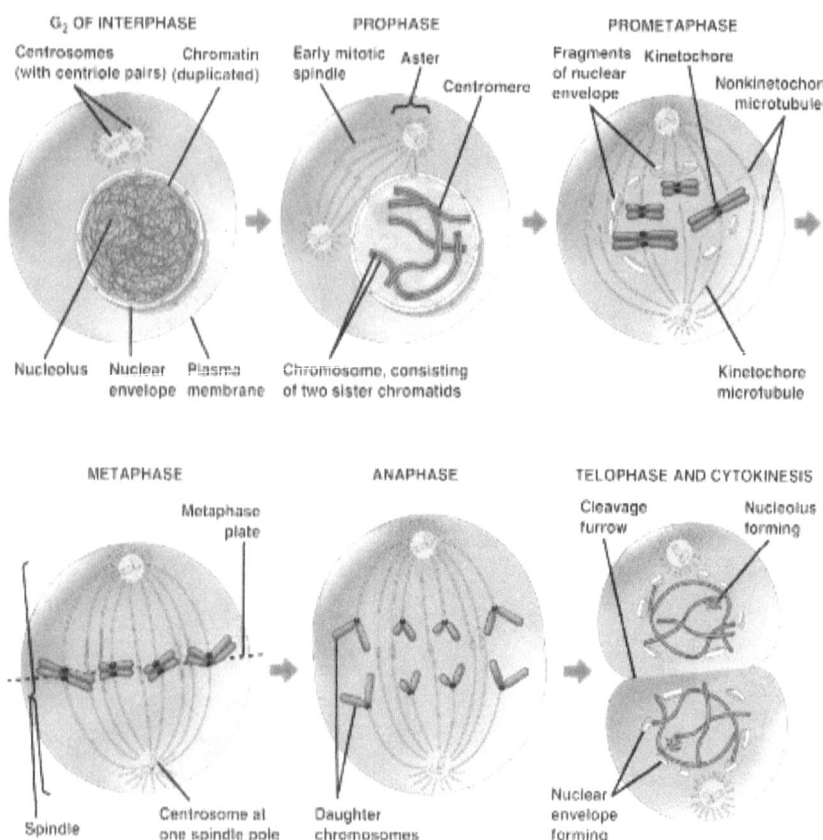

G₂ OF INTERPHASE — Centrosomes (with centriole pairs), Chromatin (duplicated), Nucleolus, Nuclear envelope, Plasma membrane

PROPHASE — Early mitotic spindle, Aster, Centromere, Chromosome, consisting of two sister chromatids

PROMETAPHASE — Fragments of nuclear envelope, Kinetochore, Nonkinetochore microtubules, Kinetochore microtubule

METAPHASE — Metaphase plate, Spindle, Centrosome at one spindle pole

ANAPHASE — Daughter chromosomes

TELOPHASE AND CYTOKINESIS — Cleavage furrow, Nucleolus forming, Nuclear envelope forming

Single-celled organisms reproduce asexually through binary fission. They divide in half.

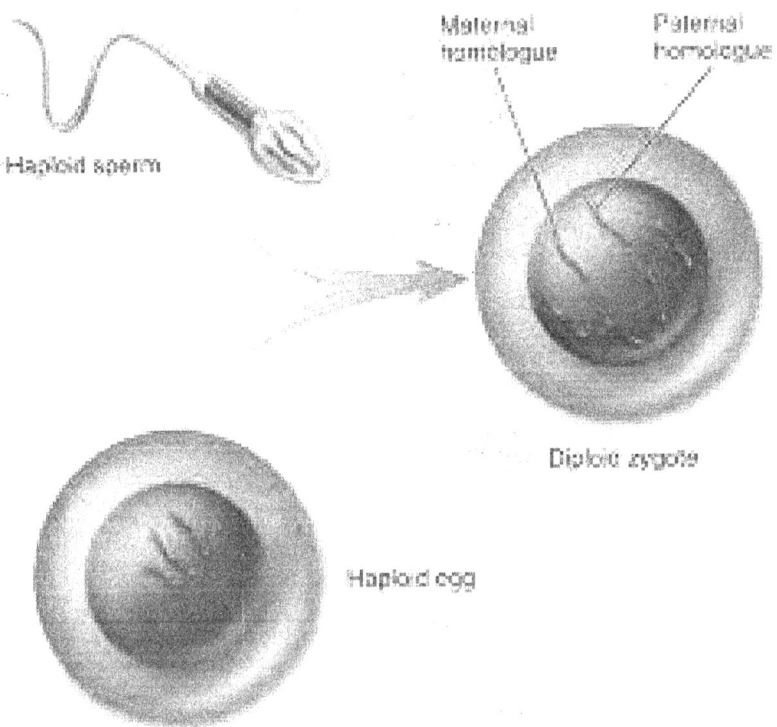

Sperm, Ova, and Zygote.

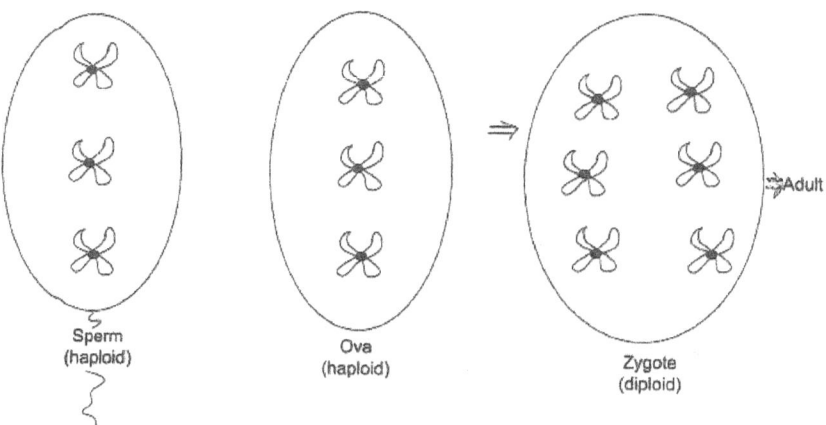

Mature Cells and Daughter Cells.

Development Anatomy

Let us talk about the origin of cells. Where do they come from? How do they know what they are supposed to become and what they are supposed to do?

Cells develop from the morula. When a sperm unites with an ovum, it forms a zygote. After a few cell divisions, the zygote becomes a morula, which means "mulberry" in Latin.

The DNA in each of our cells contains the blueprint for our entire body. Morulas take advantage of this. They contain a high concentration of stem cells, which are totipotent and can become any cell in the body. What decides what they will become are electrochemical signals that they receive from surrounding cells.

The morula has a primitive gut and an anal pore. It is surrounded by the following three layers of tissue:

> Endoderm—Latin for "inner skin." It forms the lining of the heart, arteries, capillaries, and veins.

> Mesoderm—Latin for "middle skin." It forms muscles, organs, bones, tendons, blood vessels, and lymphatics.

> Ectoderm—Latin for "outer skin." It will become skin, the neurons of the central nervous system, and the lining of hollow viscera.

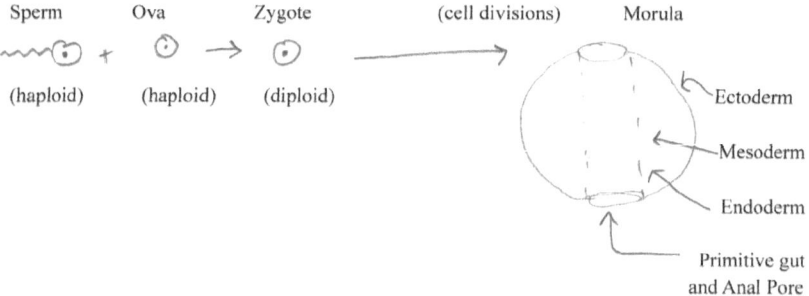

Sperm Ova Zygote (cell divisions) Morula
(haploid) (haploid) (diploid)

Ectoderm
Mesoderm
Endoderm
Primitive gut and Anal Pore

Biologists learned the above by injecting morulas with radioactive isotopes and then using scintillation wells to see where they ended up.

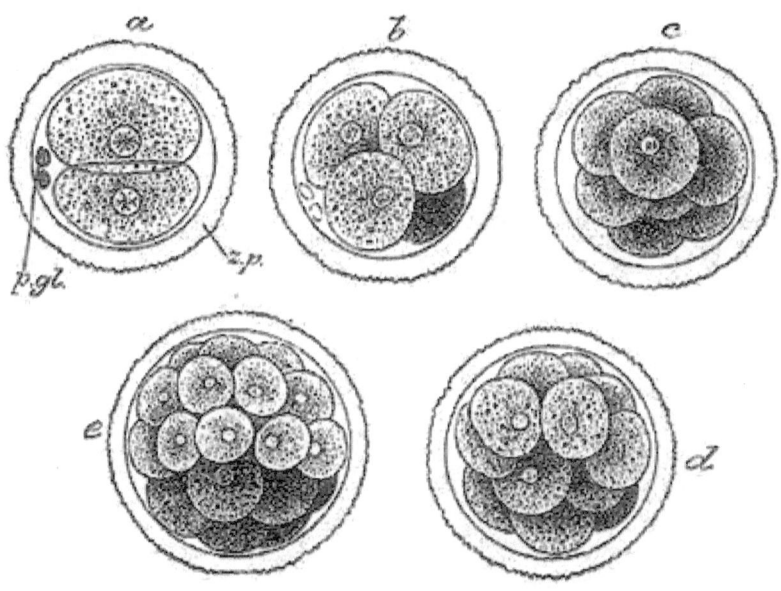

Zygote undergoes cell division.

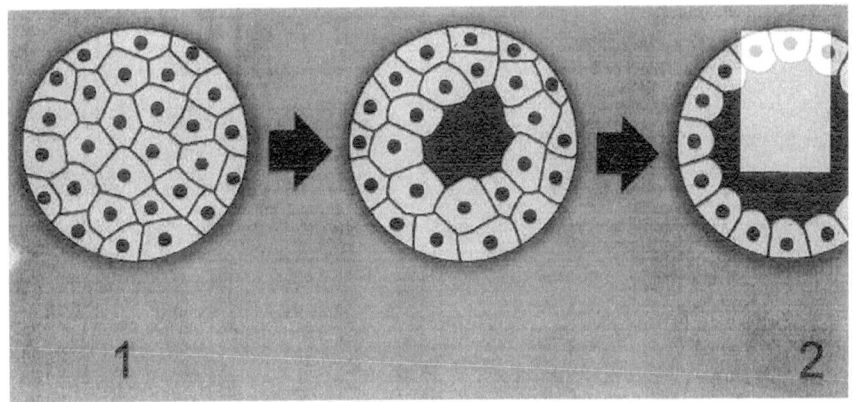

Morula.

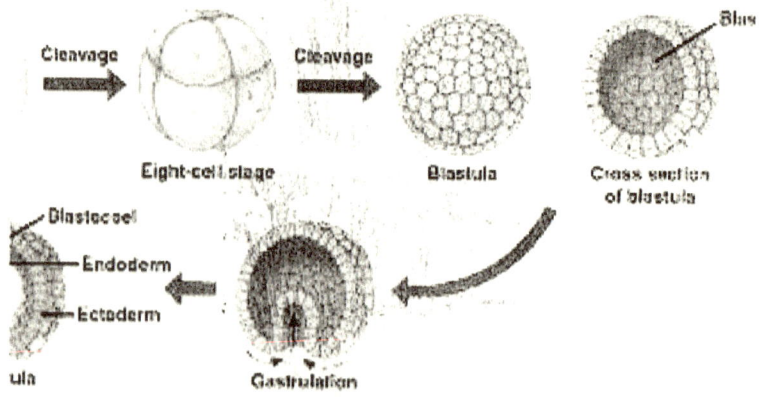

In the resting state, the nucleus of a cell contains a single chromosome that is coiled upon itself like a ball of yarn. It contains the blueprint for the entire body and is long enough to reach from the Earth to the sun (92,505,000 miles). It tells the cell what to become, how to function, and how to reproduce. It also tells cells when to die—a process known as cell apoptosis. This protects it against accumulating genetic mistakes and becoming cancerous.

Totipotency refers to the ability of stem cells to differentiate into any cell in the body.

Ontogeny Recapitulates Phylogeny' means that when an embryo develops in the womb it resembles it resembles its primitive ancestors. First it looks like an amoeba. Then a jelly fishes. Then a fish, a lizard, a mammal, and finally a human being.

A tissue is a group of similar cells that performs a common function. An organ is a collection of tissues.

During the first trimester of pregnancy (the first three months), electrochemical messengers tell the cells of the morula what to become. It is a process known as organogenesis, and each organ can form only once. It is the most critical part of pregnancy. Most women do not know that they are pregnant until organogenesis is half over (six weeks after conception).

Let us take another look at the cells of the human body.

Animal cells contain a

1. Nucleolus, which lies at the center of the nucleus and produces ribosomal RNA.
2. Chromosomes, which are in the nucleus. They direct the activity of the cell and make it possible for cells to divide.

3. Nuclear membrane, which surrounds the nucleus and makes it possible for certain substances to enter and leave.
4. Rough Endoplasmic Reticulum (RER), which synthesizes proteins.
5. Lysosomes, which contain a caustic enzyme called lysozyme. It digests foreign invaders and nutrients.
6. Vacuoles, which transport water, nutrients, proteins, and wastes.
7. Golgi bodies, which transport lipids.
8. Smooth Endoplasmic Reticulum (SER), which synthesizes lipids.
9. Mitochondria, which provide cells with energy in the form of ATP (adenosine triphosphate). Most plant cells contain 400 to 500 mitochondria. Most animal cells contain 4,000 to 5,000.
10. Cell membrane, which surrounds the cell and selectively allows certain substances to enter and leave.

Anatomy of A Cell

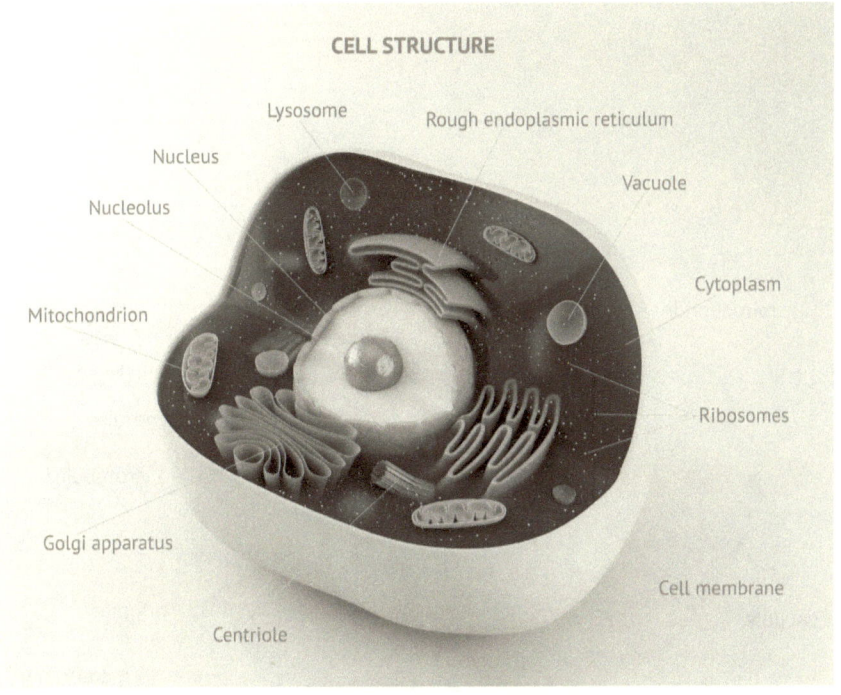

CELL STRUCTURE

Lysosome

Rough endoplasmic reticulum

Nucleus

Vacuole

Nucleolus

Cytoplasm

Mitochondrion

Ribosomes

Golgi apparatus

Cell membrane

Centriole

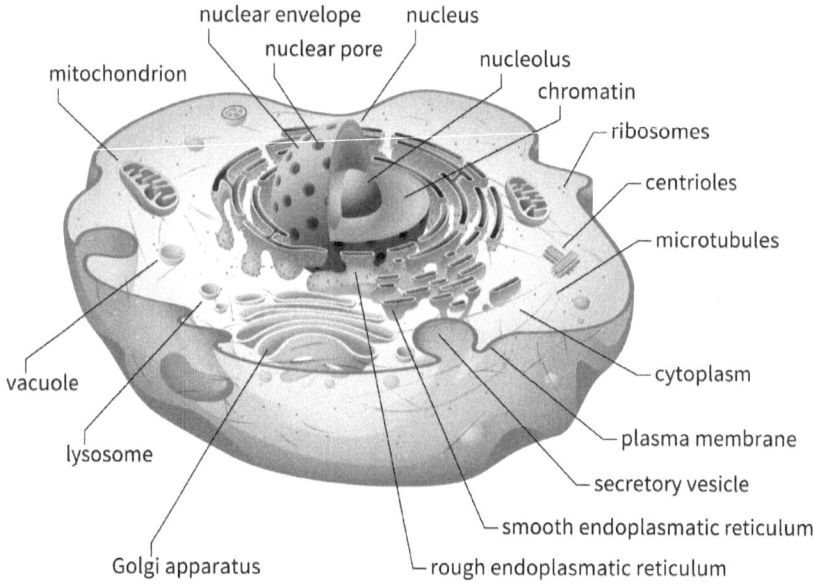

Animal cell.

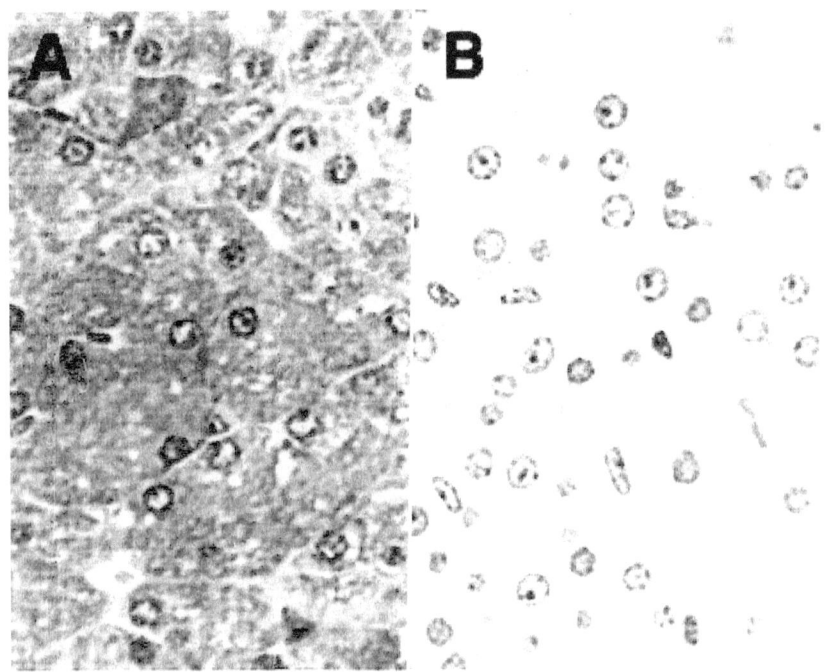

Cells with Giemsa staining (organelles).

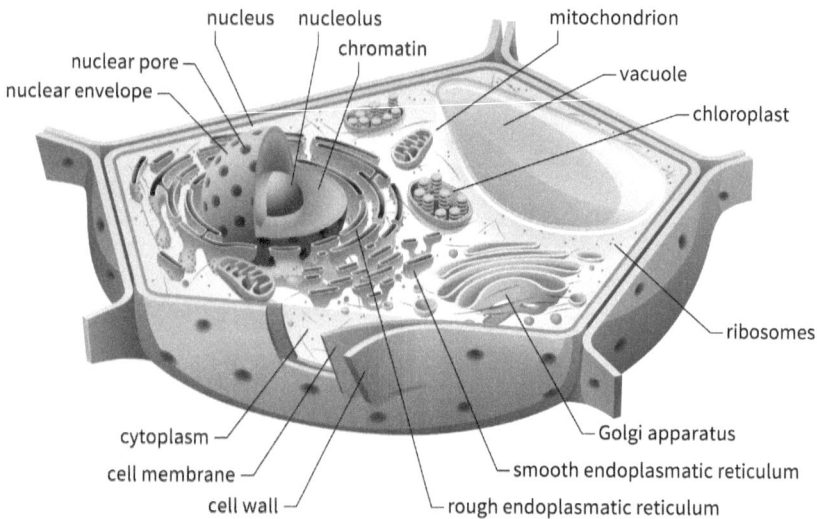

Plant cell.

This is what cells look like. Let us study them in more detail. Plant cells contain the same structures as animal cells and two more:

1. Cell walls that give them rigidity.
2. Chloroplasts that give them green color and synthesize oxygen and food through photosynthesis.

Cells can grow, synthesize special substances, generate ionic currents, and reproduce.

Cells consist of the following:

- A cell membrane that surrounds them.

  Cytoplasm (a gelatinous substance within them).

  Nucleus (directs the activity of cells and allows them to reproduce).

  Organelles (microscopic structures that perform many functions).

Cells of The Human Body Include

Epithelial cells, also known as squamous cells (which means "tile-like" in Latin). They arise from the epithelial layer of the morula and give rise to skin, neurons, and the lining of hollow viscera.

Striated cells, which are voluntary muscles that arise from mesoderm, such as muscles of the hands.

Smooth muscles, which also arise from mesoderm and form involuntary muscles such as the muscles of respiration.

Cardiac muscles, which arise from mesoderm and drive the heart.

Epithelial cells.

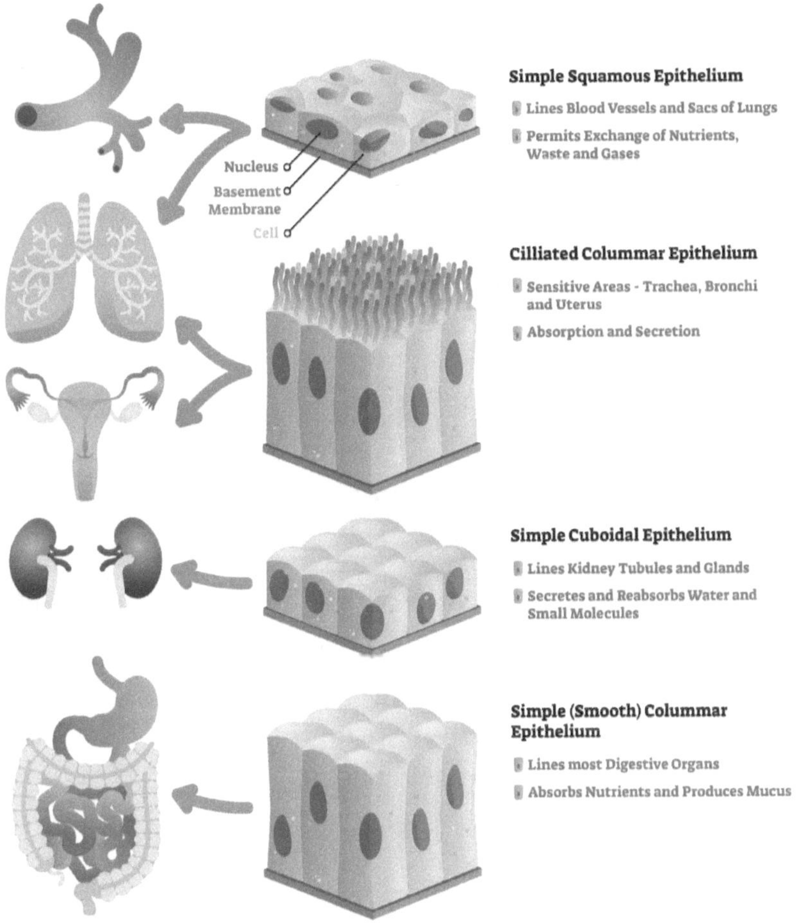

Simple Squamous Epithelium

- Lines Blood Vessels and Sacs of Lungs
- Permits Exchange of Nutrients, Waste and Gases

Nucleus

Basement Membrane

Cell

Cilliated Colummar Epithelium

- Sensitive Areas - Trachea, Bronchi and Uterus
- Absorption and Secretion

Simple Cuboidal Epithelium

- Lines Kidney Tubules and Glands
- Secretes and Reabsorbs Water and Small Molecules

Simple (Smooth) Colummar Epithelium

- Lines most Digestive Organs
- Absorbs Nutrients and Produces Mucus

Types of epithelial cells are ciliated columnar, simple columnar, simple cuboidal, and simple squamous cells.

Muscle fiber

Nucleus

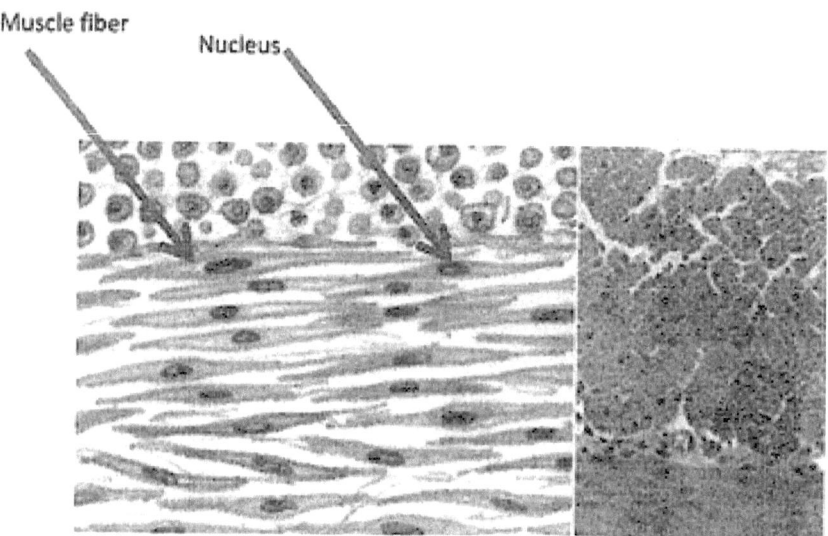

Striated muscle.

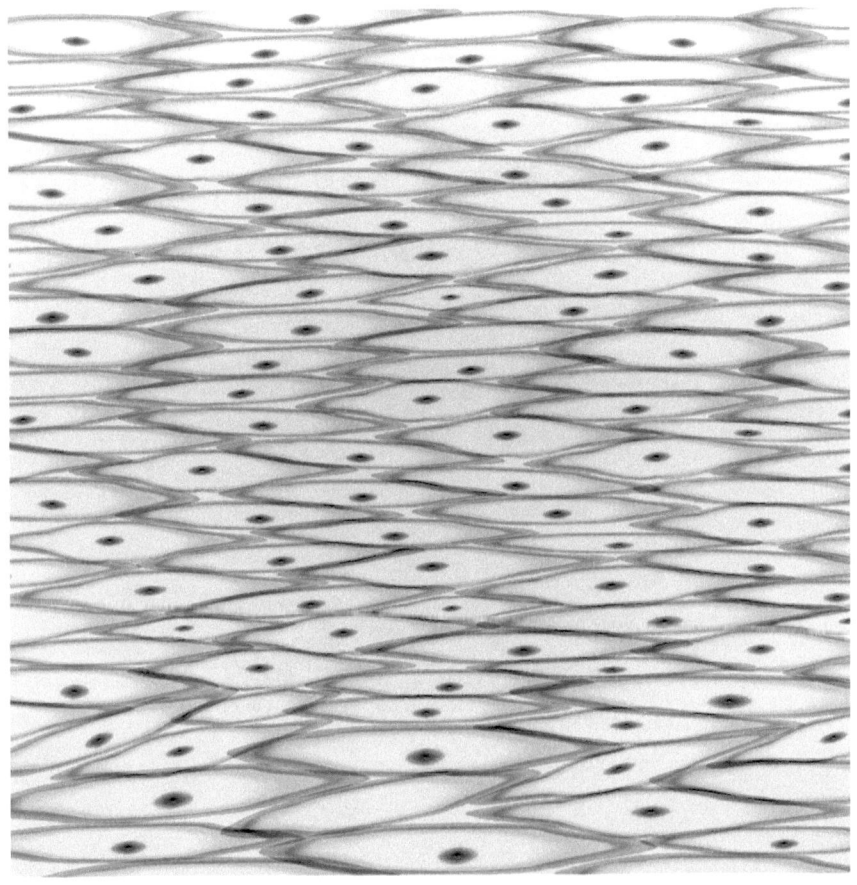

Smooth muscle tissue.

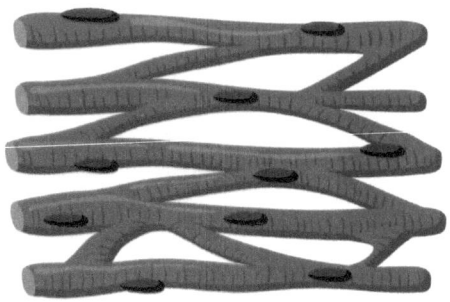

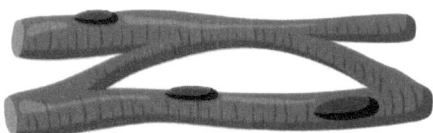

Cardiac muscle.

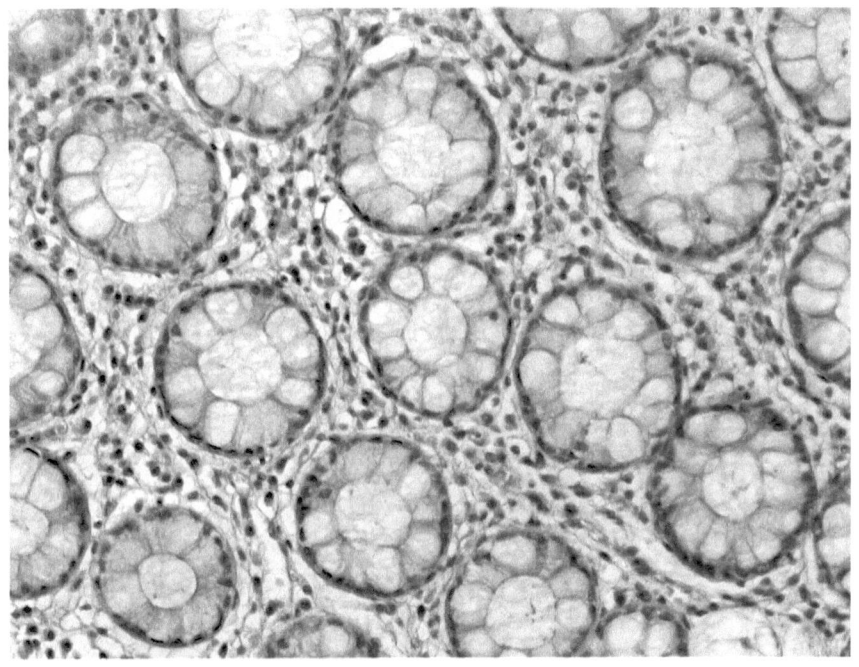

Glandular tissue.

The Skeletal System

Let us move on to the skeletal system. We will begin with the Haversian system of the long bones. If we make a transverse cut through a long bone, we will see the Haversian system. It would look like a tree stump. It would have rings, each representing one year of life. The center would contain marrow, which once made the formed elements of the blood: platelets, red blood cells, and white blood cells.

During childhood the bones would have contained red marrow. It would have been red, and 2 percent of its cells would have been totipotent. They could have become anything, but they transformed into the formed elements through a process known as megakaryocytosis (which creats platelets) and the formation of red and while blood cells.

As we move away from the center, we will come across nerve fibers that innervate the bone, and blood vessels.

The outermost layer would be the cortex, which is hard and is surrounded by a neural net called the periosteum.

In children the ends of long bones contain epiphyseal plates that allow them to grow and remain soft until the end of puberty (approximately age seventeen); then they "seal" and become hard. Prior to sealing the ends, the bones receive blood vessels from vessels within the bone and from branches that extend from the adjacent bone. But post puberty the branch that traverses the epiphyseal plate atrophies, and the blood comes entirely from the opposite bone.

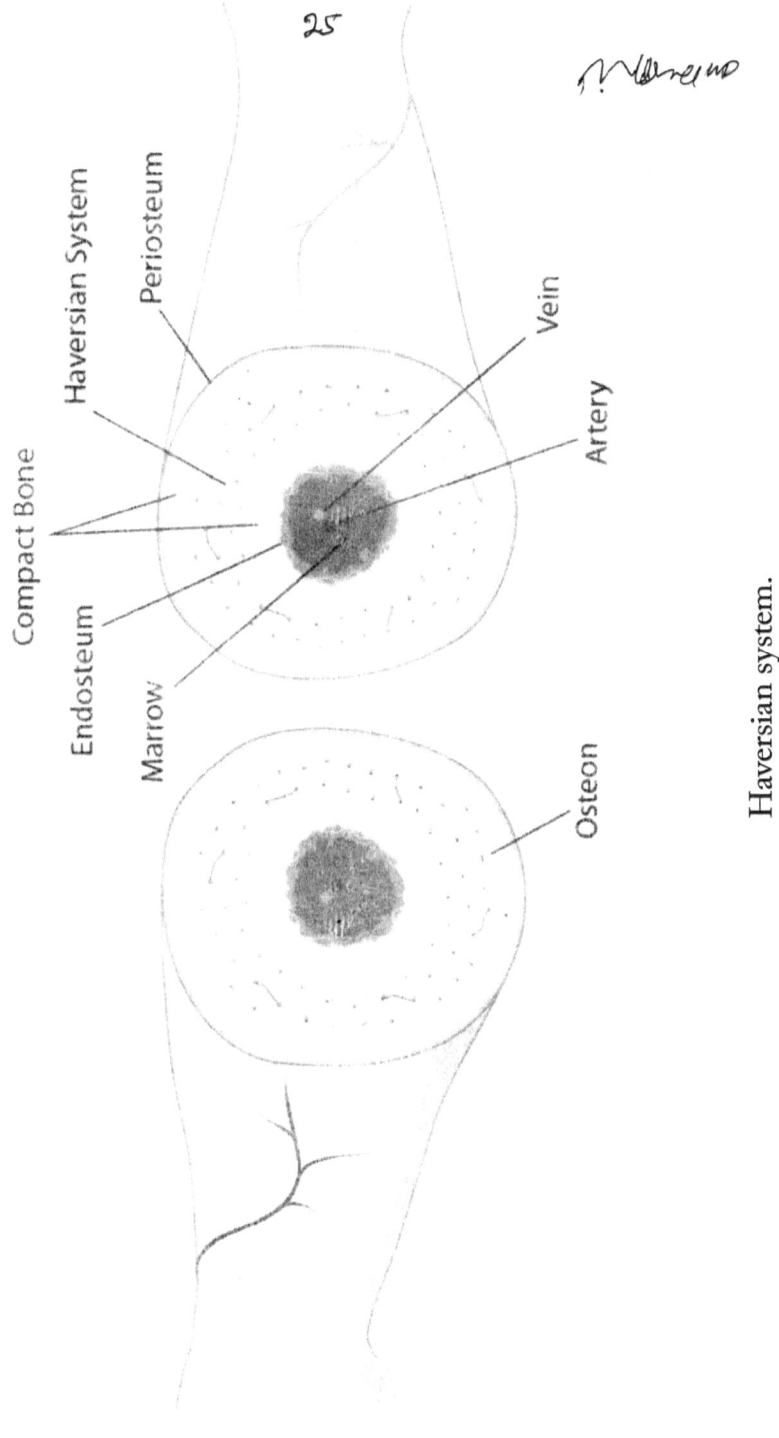

Haversian system.

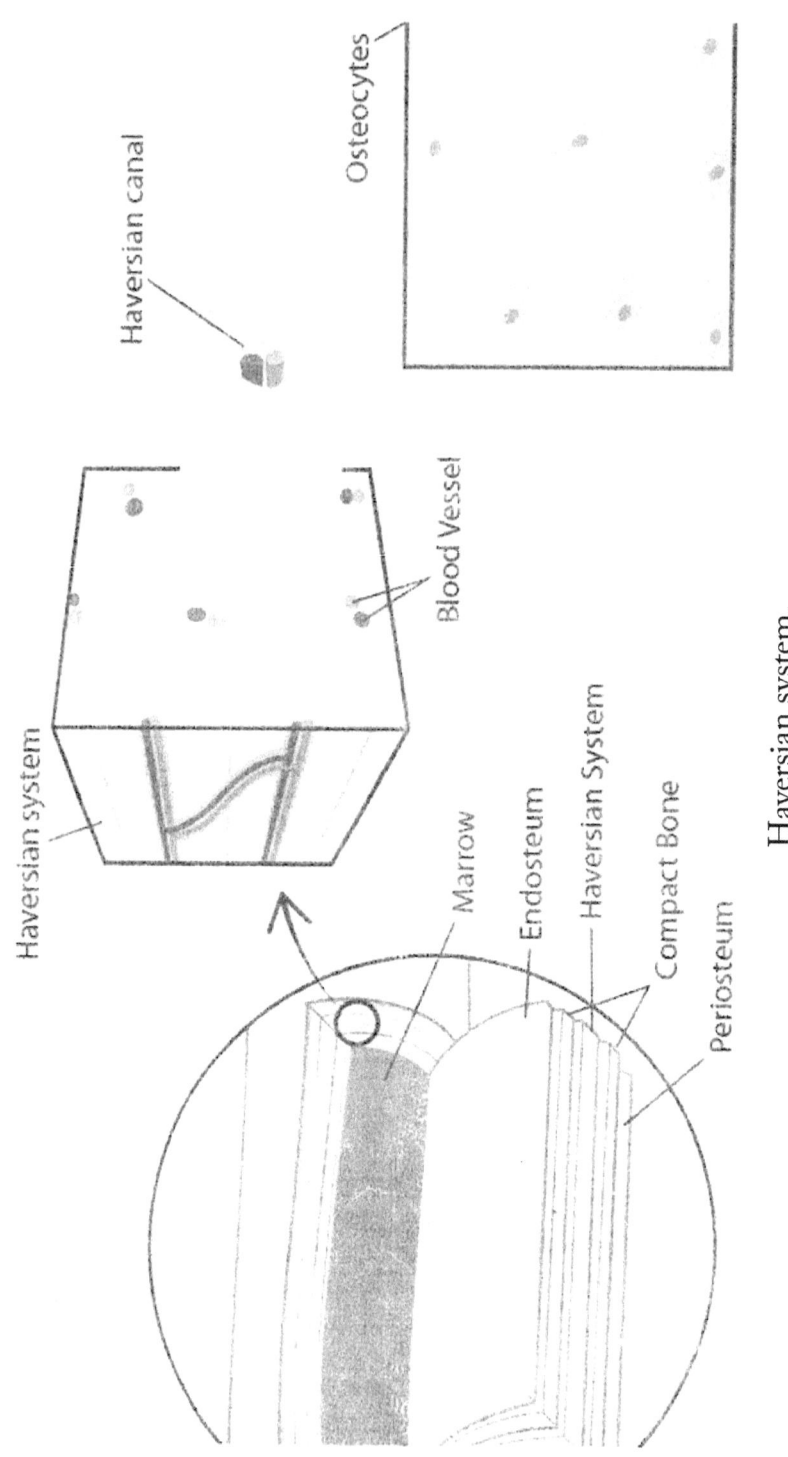

Haversian system.

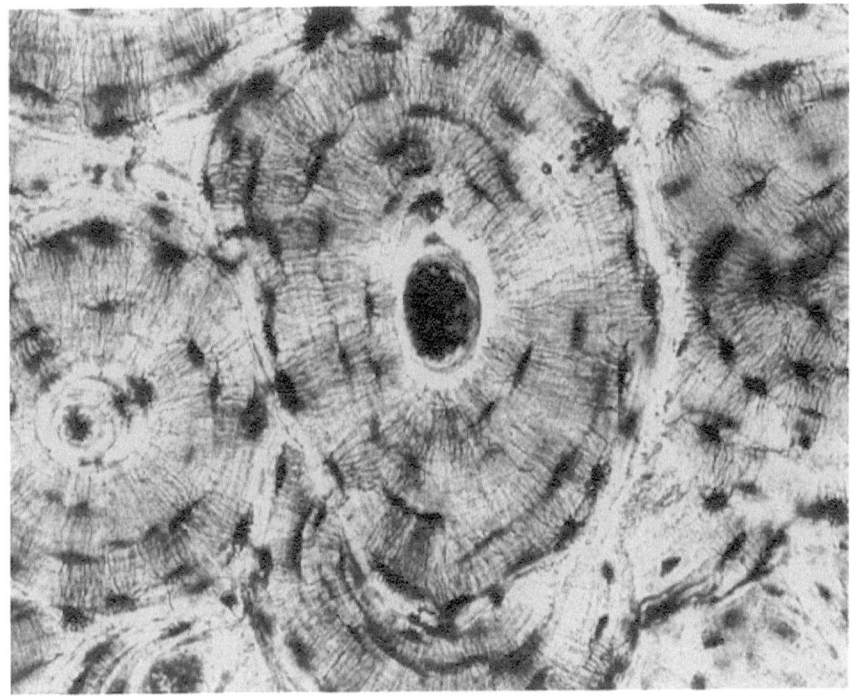

Haversian system.

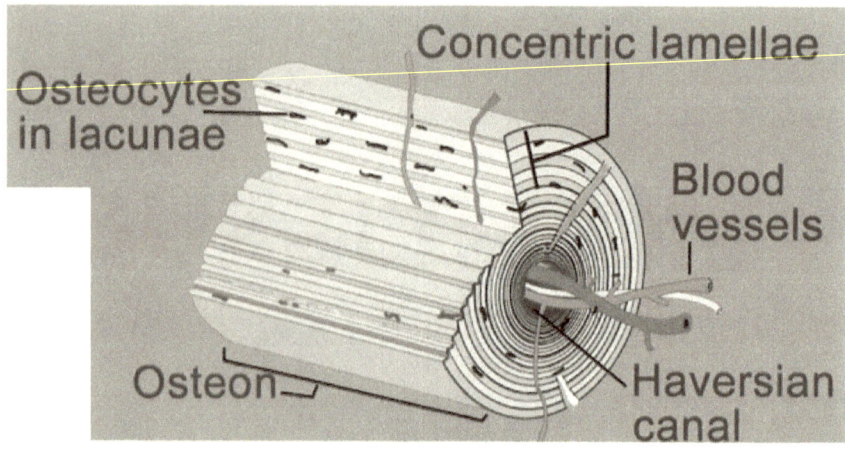

Cross-Section of a Haversian system.

Gastrointestinal Tract

Now let us move on to the intestines.

The small intestine begins at the distal end of the stomach and extends to the caecum, near the pelvic brim on the right. It is 10 meters in length and consists of the:

1. duodenum
2. jejunum
3. ilium

The ampulla of Vater creates an orifice that opens on the side of the duodenum. It pumps a green liquid into the duodenum called bile. It contains sodium bicarbonate and a mixture of lipase and emulsifying agents that are activated by high pH.

Bile is produced by the liver and travels through the hepatic duct, which joins the cystic duct (that comes from the gallbladder). The two ducts form the hepatocystic duct, which drains into the ampulla of Vater and into the duodenum.

Emulsifying agents break lipids down into small particles. Then the lipase and sodium bicarbonate break them down further.

The outer surface of the small intestine is the lamina propria. The wall between the lamina propria and the lumen contains Peyer's patches, which assist in digestion. The villi (meaning "fingers" in Latin) increase the surface area of the lumen. Every part of the small intestine is vital, and surgeons must remove as little as possible when performing bypass surgery.

Small Intestine:

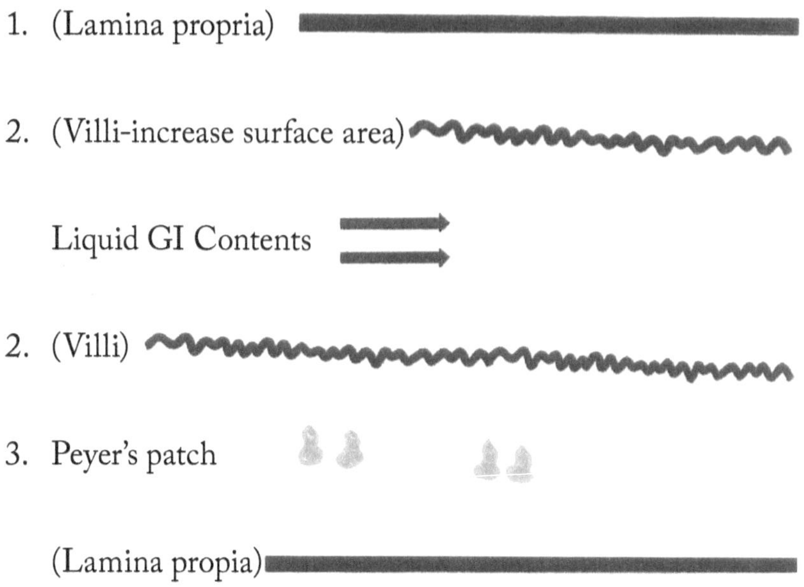

1. (Lamina propria)

2. (Villi-increase surface area)

Liquid GI Contents

2. (Villi)

3. Peyer's patch

(Lamina propia)

1. Lamina propria (outer surface)
2. Villi (inner surface)
3. Peyer's patches (aid in digestion)

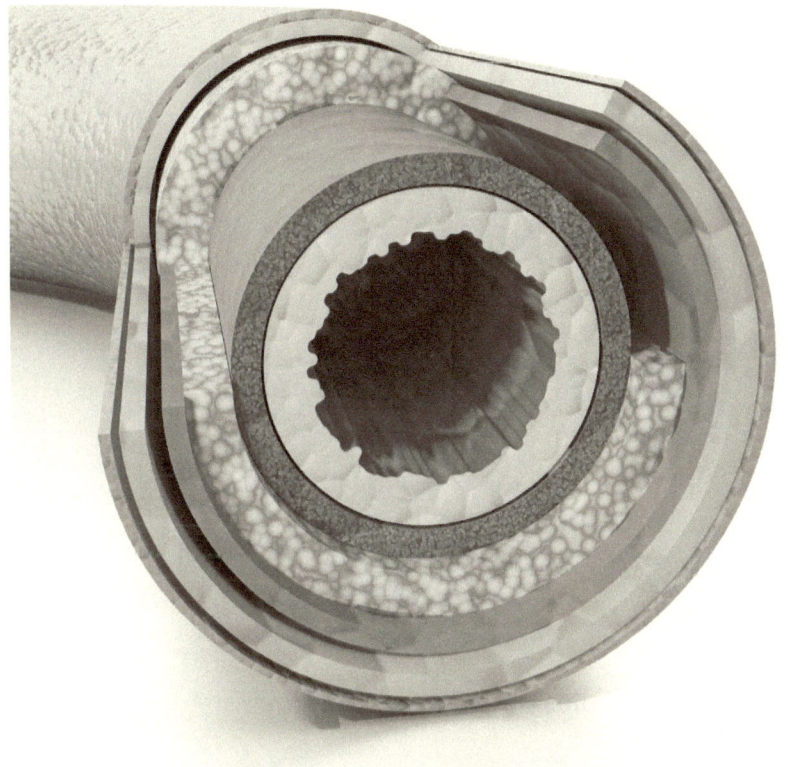

GI Tract

**INTESTINAL
EPITHELIAL CELL**   **INTESTINAL VILLI**

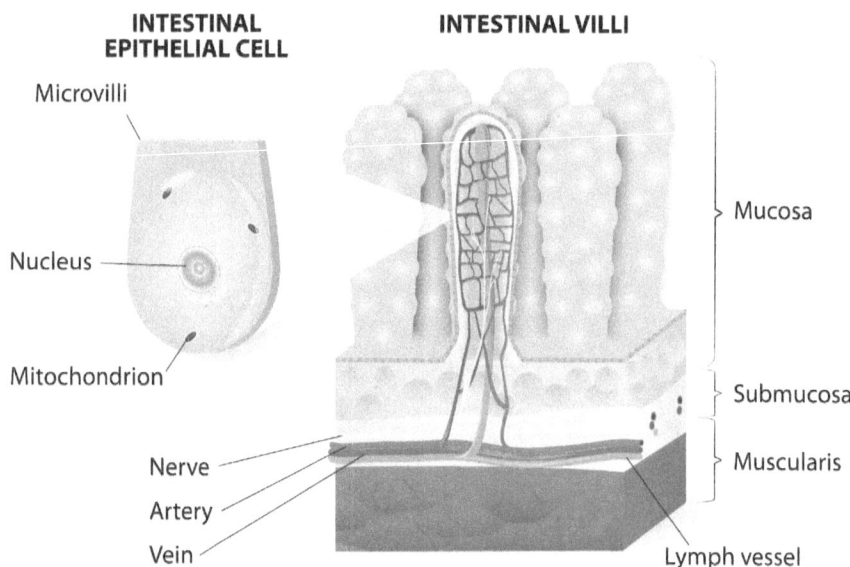

Microvilli

Nucleus

Mitochondrion

Nerve

Artery

Vein

Mucosa

Submucosa

Muscularis

Lymph vessel

Villus of the Small Intestine:

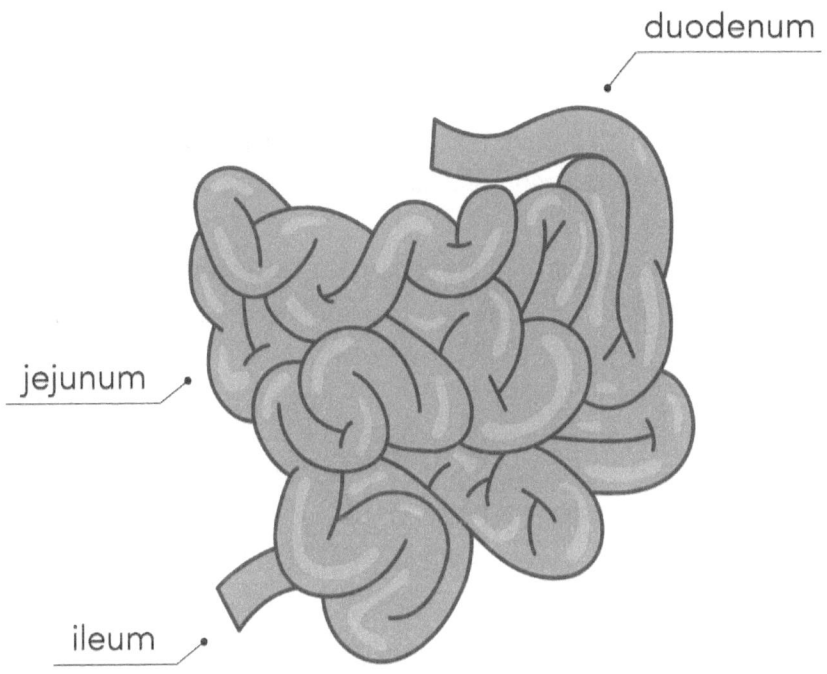

duodenum

jejunum

ileum

Small intestine.

The Central Nervous System

Let us move on to the neurons that make up the central nervous system. They arise from ectoderm and consist of:

1. Dendrites, which means "branches" in Latin. They are stimulated by neurotransmitters, which they receive from other neurons.
2. Soma, which means "body." It contains the nucleus.
3. Axon, which means "tail." It transmits impulses to the axon terminal and has side ports that are influenced by other neurons.
4. Axon terminal, which transmits neurotransmitters (such as acetylcholine and epinephrine) across the synaptic gap to the dendrites of other neurons.

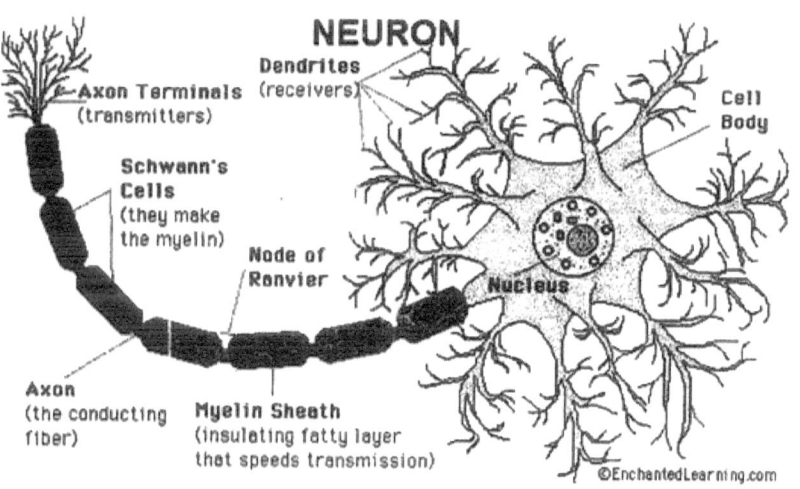

Lungs

Lungs arise from mesoderm, but the lining of the air passages comes from ectoderm. Bronchi and bronchioles are conduits that allow air to enter the alveoli—a word meaning "bunches of grapes" in Latin. They consist of walls that are 1/10,000 of an inch thick and allow oxygen and carbon dioxide to diffuse into and out of capillaries that circulate blood.

During early development, lung buds grow out of the mediastinum in the area of the main bronchi. When they touch ribs and the diaphragm, electrochemical messengers cause "contact inhibition." and two things happen:

- The lungs stop growing.

    They differentiate into their adult form.

A former member of the Italian senate, Rita Levi-Montalcini, won a Nobel Prize for her work on electrochemical messengers.

The respiratory system consists of the following:

1. Trachea
2. Corina
3. Bronchi
4. Bronchioles
5. Alveoli

On the left there is an inferior and a superior bronchus.

On the right there is an inferior, a middle, and a superior bronchus.

LUNGS ANATOMY

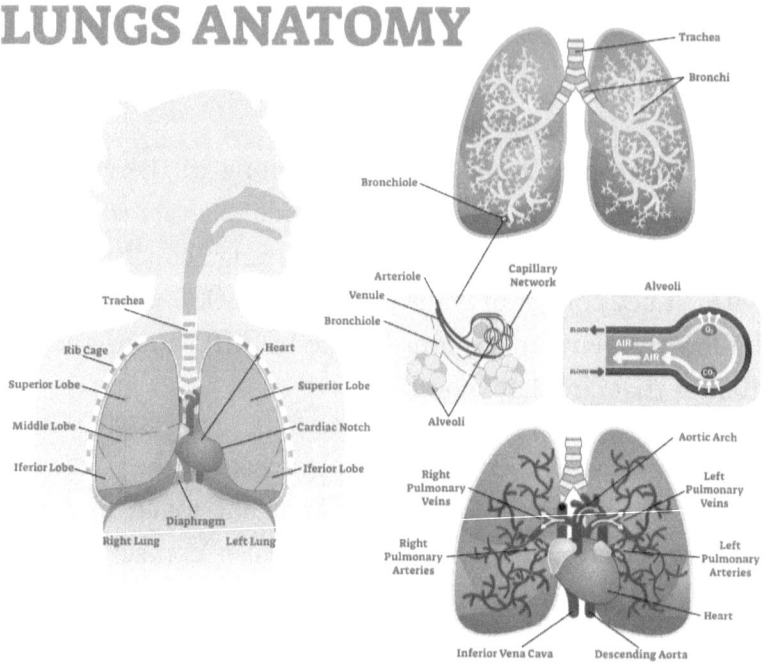

Lung.

Glandular tissue arises from mesoderm.
Bone arises from mesoderm.
Gastrointestinal systems arise from mesoderm.
Neurons arise from ectoderm.
Lungs arise from mesoderm.

Let us discuss the structure of cells. We will start with the cell membrane. It is a phospho-phospholipid-lipid bilayer that contains pores. They allow substances to enter and leave selectively. If we look at the membrane with X-ray crystallography, this is what we see:

Cell Membranes

Consist of a phospho-phospholipid lipid bilayer and three different kinds of pores.

1. Phosphate side chain
2. Lipid
3. Space
4. Second lipid molecule
5. Phosphate

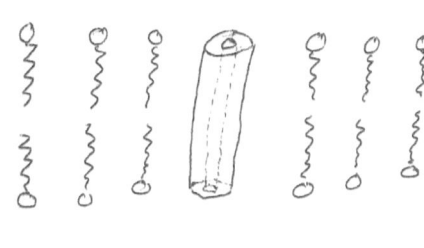

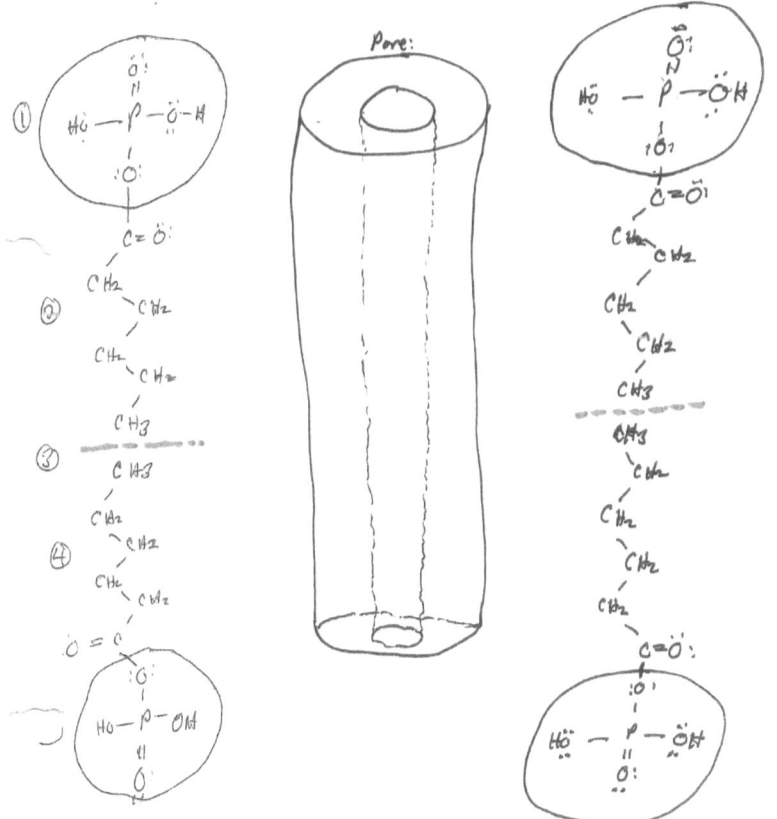

Cell membranes.

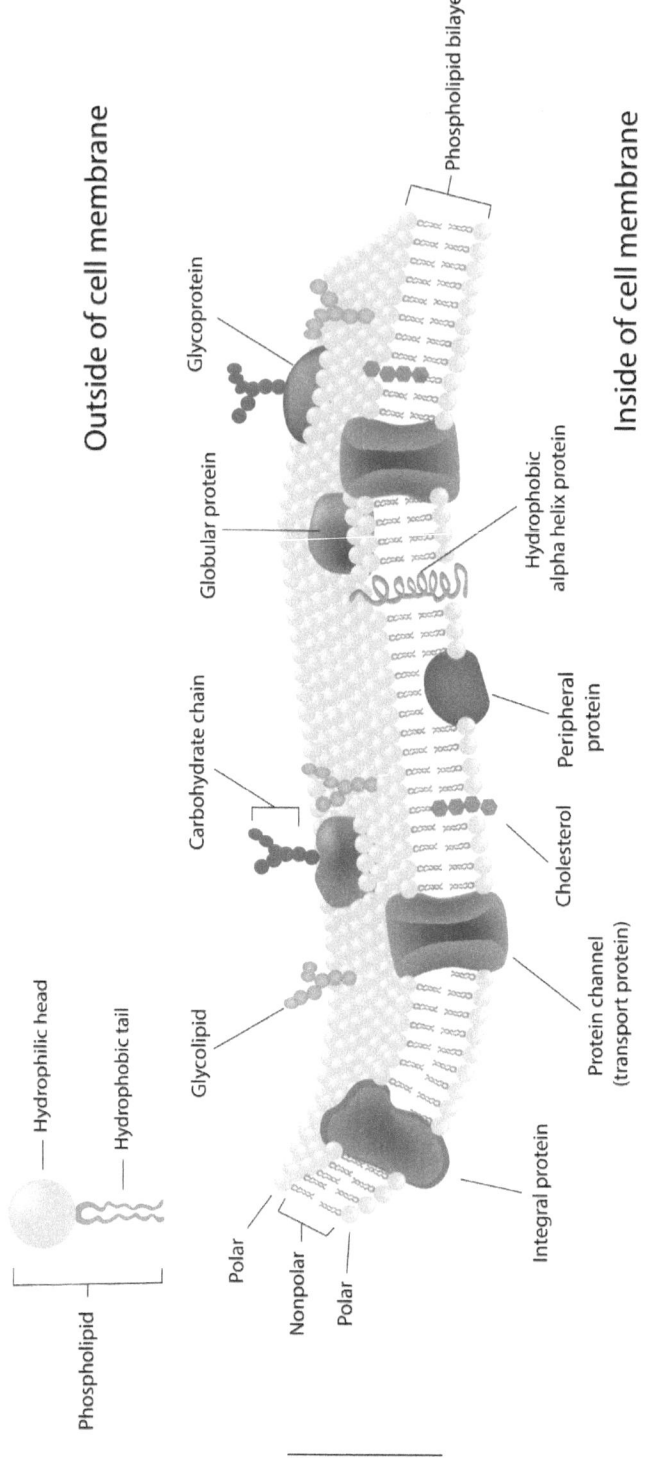

Cell membrane.

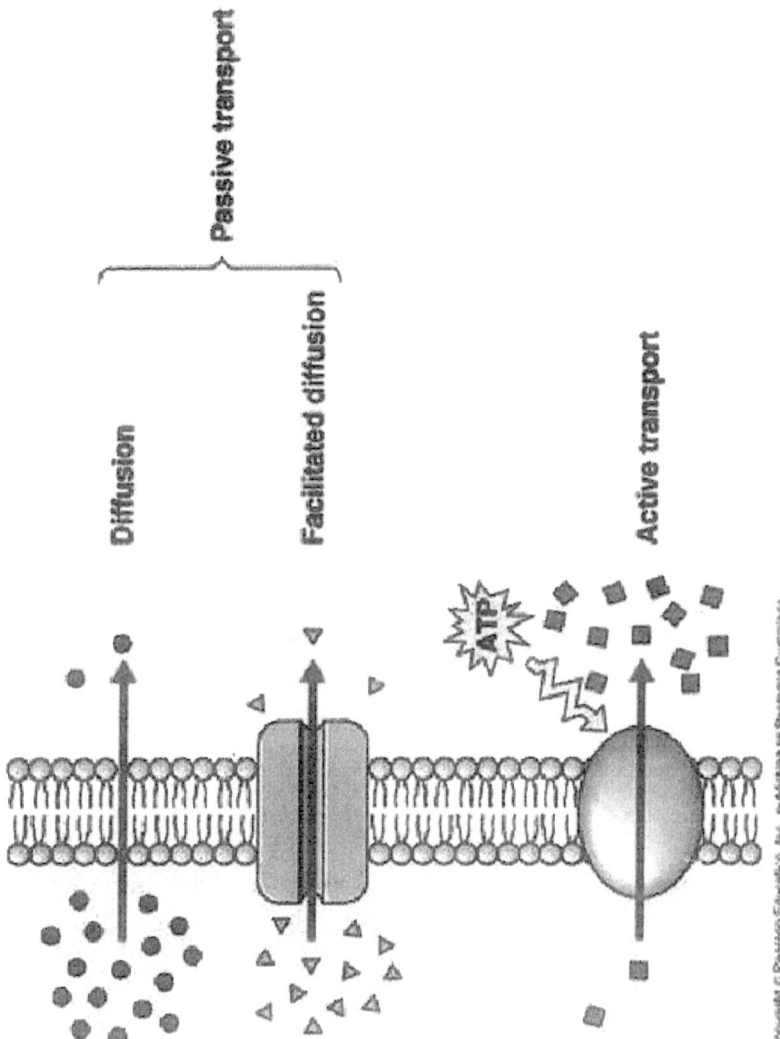

Pores in cell membrane.

As mentioned, the membrane contains three kinds of pores:

Passive Perfusion	Facilitated Transport	Active Transport
(where no energy is required)	(one thing goes in and something else comes out)	(energy is required to move substances in or out of the cell, and it comes from ATP)

 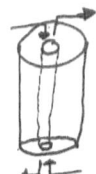 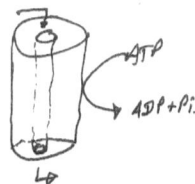

Membrane

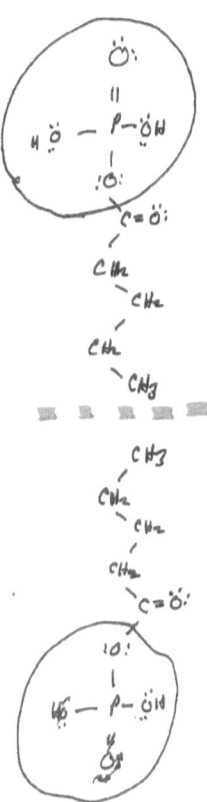

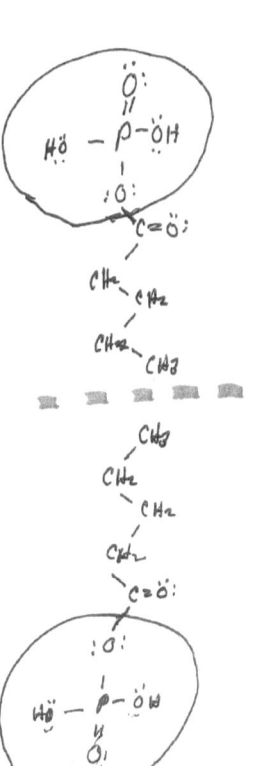

There are four ways for substances to enter cells:

1. They could diffuse through the lipid-lipid bilayer directly.
2. They could pass through one of three different kinds of pores in the cell membrane.
3. A cell could consume a solid particle through phagocytosis.
4. A cell could consume a liquid with pinocytosis.

Once inside the cell, substances would be packaged into a Golgi body or a vacuole and would move under the influence of a microtubule.

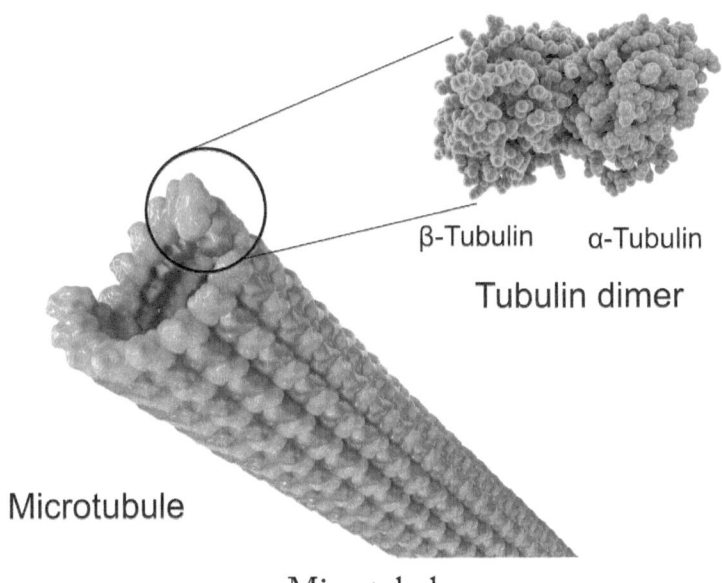

β-Tubulin   α-Tubulin

Tubulin dimer

Microtubule

Microtubule.

Now that we understand how cells consume outside substances, let's think about the opposite. How do they make and secrete substances?

Golgi bodies obtain lipids from the smooth endoplasmic reticulum. Then they migrate to the cell membrane and secrete them.

Microtubules propel Golgi bodies and vacuoles across the cell. They can work in two directions. After it obtains lipids, proteins, foreign invaders, or damaged cells at the cell membrane (through phagocyctosis and pinocytosis), an organelle can move through the cytosol under the influence of microfibers. Or if lipids or proteins are generated, microfibers can move the Golgi bodies and vacuoles to the cell membrane, and they can be excreted through reverse pinocytosis or phagocytosis.

Reverse Phagocytosis

Let us move on to a new topic and explain how cells secrete the substances that they synthesize. Particles of lipid are expelled through reverse phagocytosis. Golgi bodies obtain them from the SER. Then microfibrils move them across the cytoplasm to the cell membrane, and the cell membrane expels them.

The microfibers are tubes. The tubes are created by a protein called "tubulin." Tubulin is a water-soluble protein that produces rings when dissolved in water. The rings coalesce to form tubes, and the tubes contract similar to muscle fibers.

Tubulin:

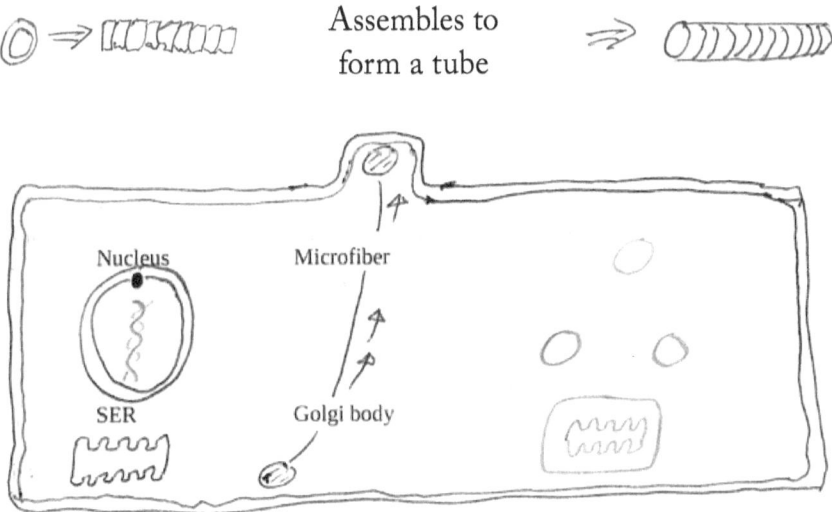

White blood cells consume:

- Fungal organisms

  Amoeba
  Protozoa
  Bacteria
  Viruses
  Cell debris
  Cancer cells

When white blood cells consume substances, they use vacuoles and lysosomes. The foreign substance enters a vacuole; then the vacuole melds with a lysosome. The lysosome contains a caustic enzyme that digests the above.

Pinocytosis

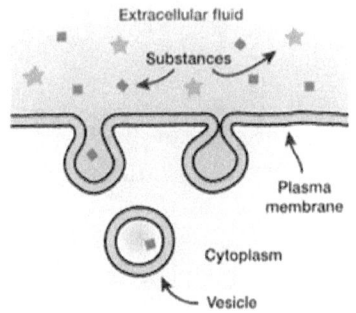

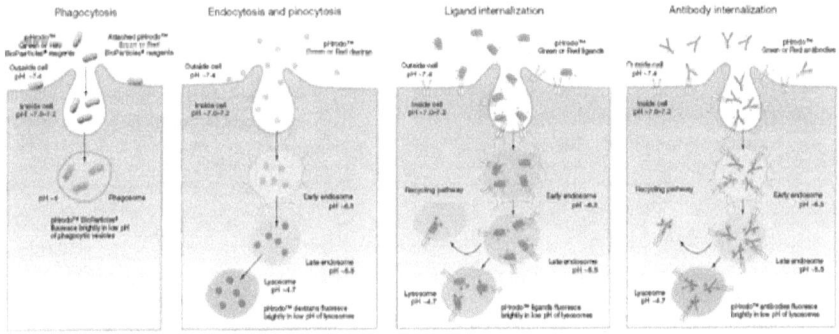

Structure of DNA

Let us move on to a discussion of DNA:

If we take an optically visible chromosome and study it with X-ray crystallography, this is what we see. This is the basic structure of DNA. We know from the image what it looks like, but we do not know what it is made of.

After additional work, Watson and Crick worked out the composition of DNA. It is a double helix (a twisted ladder) with the following structure:

1. Nucleosomes (consists of histones)
2. Deoxyribose sugar
3. Sense strand
4. Antisense strand
5. Phosphate links
6. Nitrogenous bases

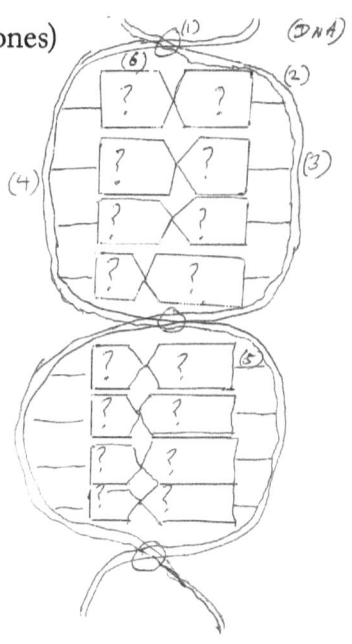

Let us talk about the ways that DNA leads to the production of certain substances such as insulin. When blood sugar is high, a chemical messenger will be released. It will interact with the nuclear proteins of DNA in the islet of Langerhan cells of the pancreas. This will cause the correct sequence of DNA to open and will expose the sense strand.

The next step is for messenger RNA to form using the exposed sense strand.

Chemical messenger: > Nucleosomes ➡ DNA Sense strand Exposed. mRNA: ⟹ sense strand m RNA:

RNA contains the same four bases as DNA with one exception. Instead of thymine we have uracil. RNA also uses ribose sugar instead of deoxyribose.

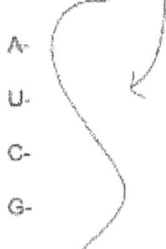

A-
U-
C-
G-

The adenine in RNA binds to thymine in DNA cytosine in RNA combines with DNA guanine, and RNA uracil combines with DNA adenine.

They combine by forming weak hydrogen bonds, 10 kcal per mole. The binding requires an enzyme called mRNA synthetase. It lowers the energy of activation needed to create the new bonds. The bonds are either double or triple, and some are stronger than others. But the average bond strength is 10 kcal, and they can break easily.

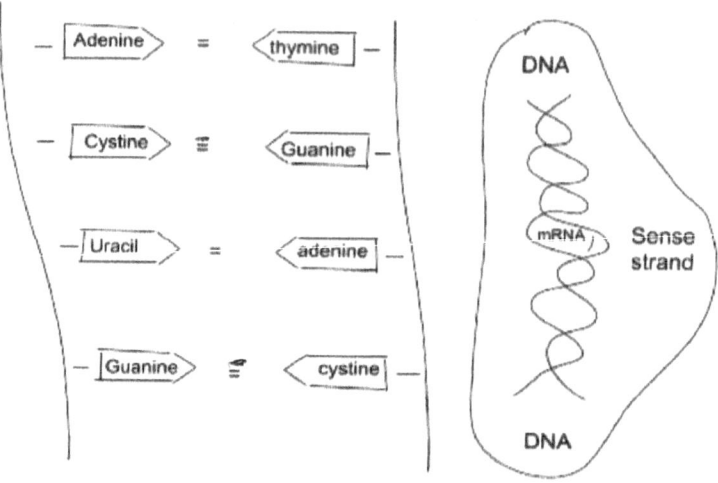

RNA-DNA

Francis Crick

James Watson

They worked out the structure of DNA.

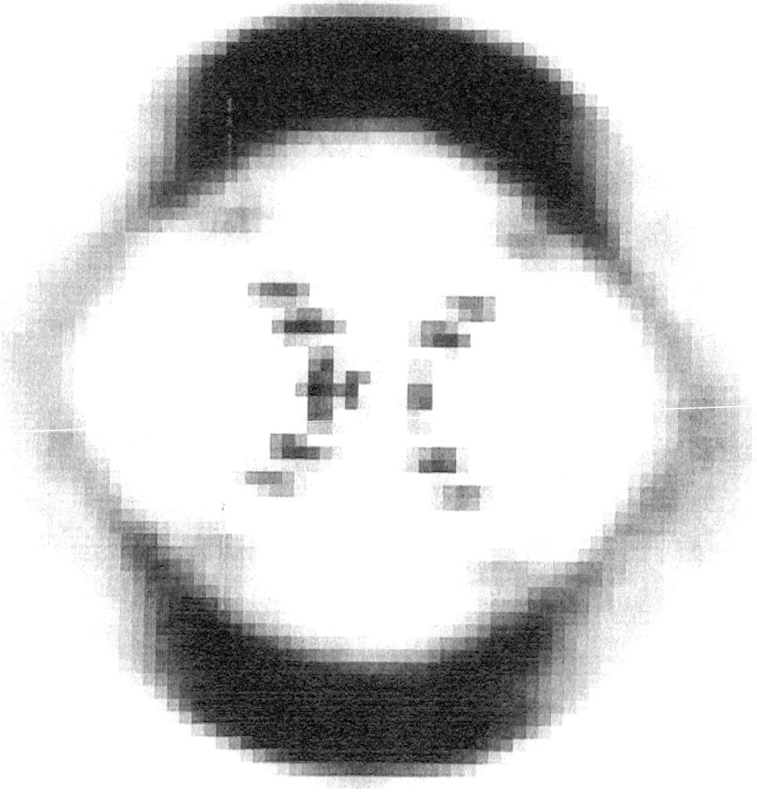

X-ray crystallography.

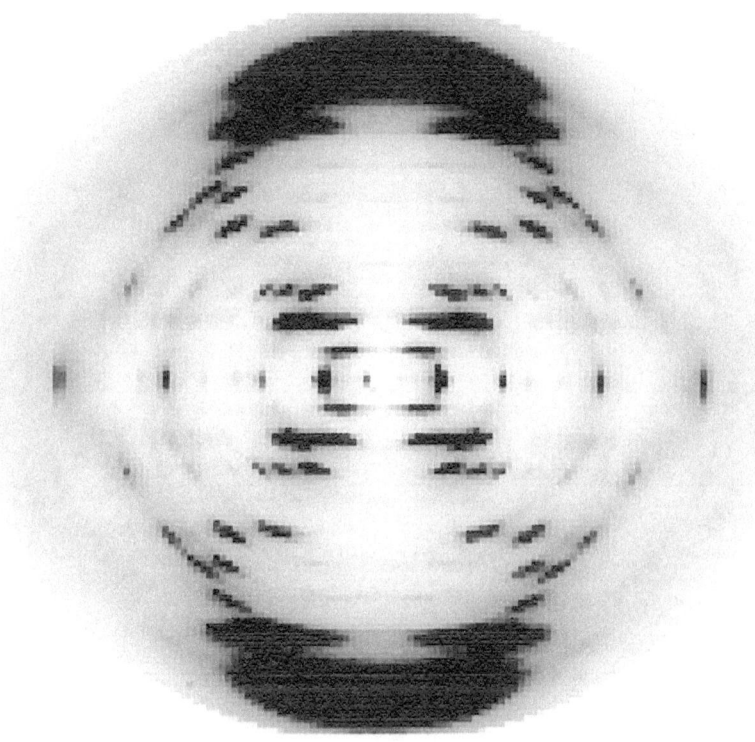

X-ray crystallography of DNA.

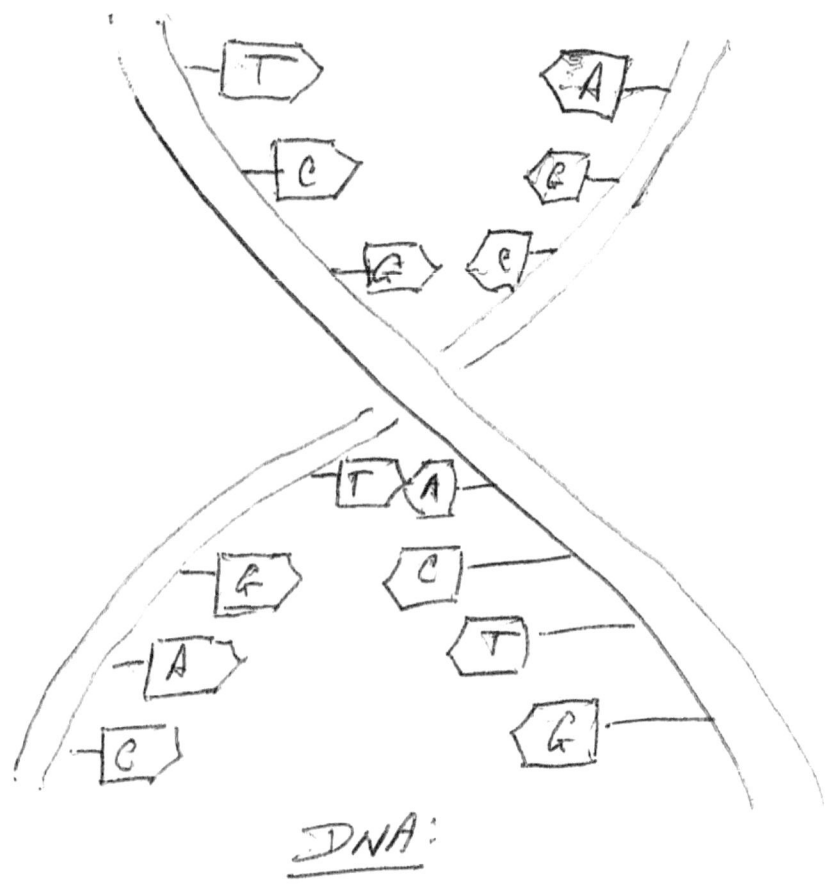

DNA:

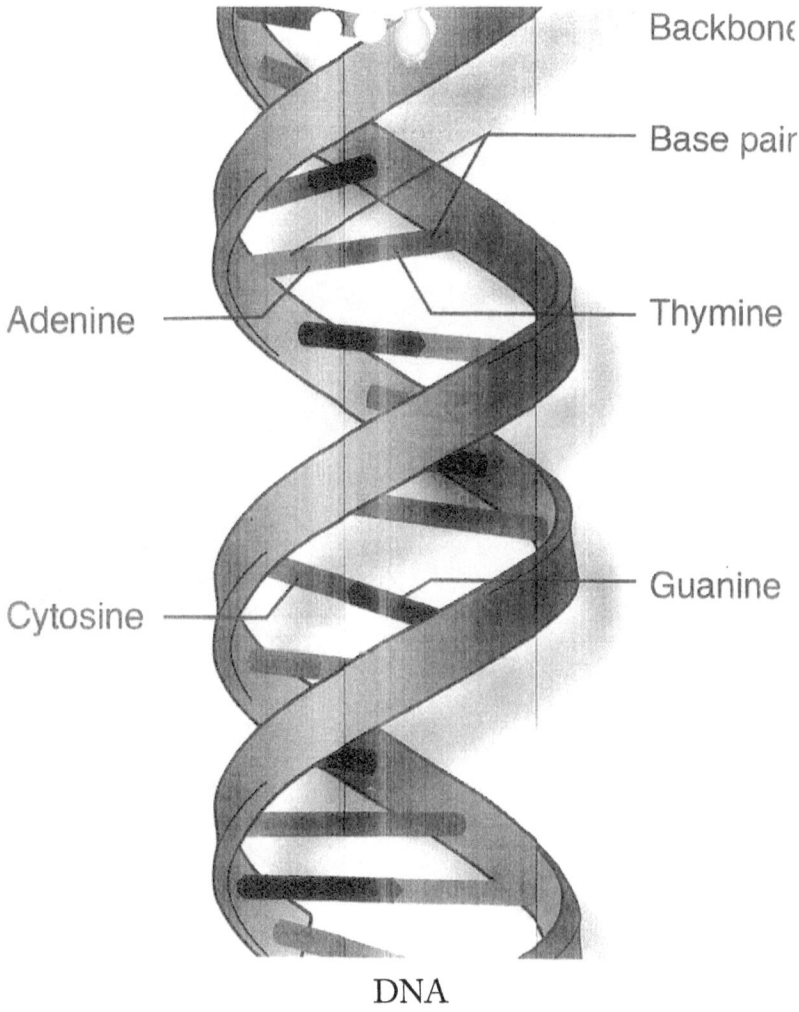

Backbone

Base pair

Adenine

Thymine

Cytosine

Guanine

DNA

Frederick Banting and Charles Best

They discovered insulin.

Once mRNA is created, the RNA polymerase will release it. Then it leaves the nucleus by passing through pores in the nuclear membrane and enters the RER. From there it interacts with ribosomes, most of which are on the walls of the RER.

Once mRNA interacts with ribosomes, it will bond with tRNA by way of a triplicate code (codon). Each unit of tRNA brings one amino acid. The orientation of the amino acids is random, and the protein polymerase positioning site has to organize them so that the C-terminal of one faces the N-terminal of another. Then the catalytic site has to lower energy of activation so that an amide bond can form. Then the final product will be released from the protein polymerase by a feedback mechanism and the protein will leave the RER and be packaged in a vacuole.

The nitrogen bases that make up the triplicate code (codon) are purines and pyrimidines. Their structures appear below:

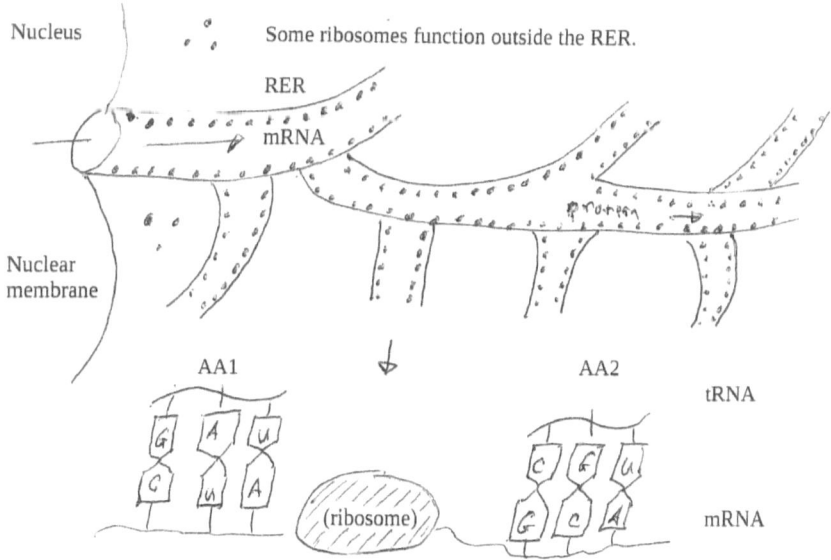

This is how a cell synthesizes proteins. Some proteins are used to build the cell. Others are enzymes that are used to promote chemical reactions, such as lipid synthesis.

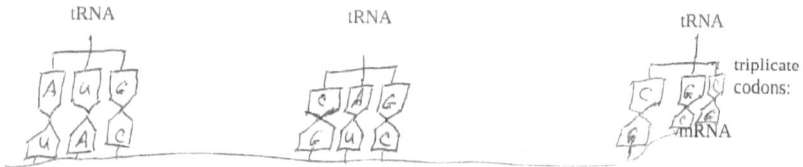

The nitrogen bases in DNA and RNA are the following:

Purines

Adenine
Guanine

Pyrimidines

Cytosine
Thymine
Uracil

Ribosome:

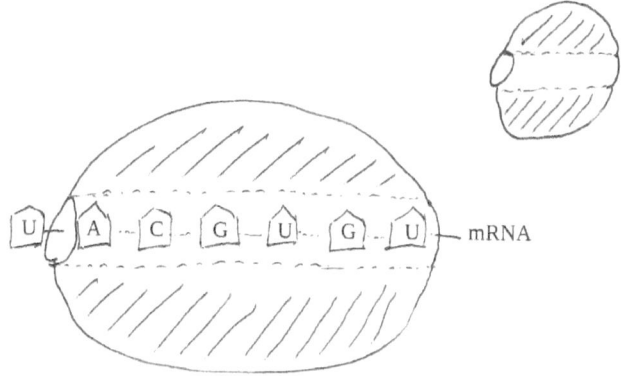

Protein molecules are synthesized by ribosomes.

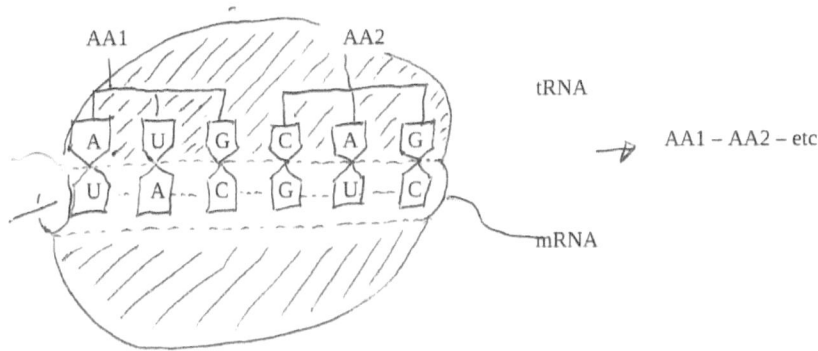

For amino acids to form proteins, they have to line up correctly. The positioning site causes the C-terminal of one amino acid to face the N-terminal of another. Each unit of transfer RNA brings one amino acid. Messenger RNA couples with transfer RNA by way of a triplicate code and determines which amino acids will be linked to form a particular protein.

To complete our understanding of this, we need to understand one more concept. To link amino acids, we are going to need protein polymerase.

As is true of all enzymes, protein polymerase has a positioning site and a catalytic site. The positioning site aligns the amino acids in the correct orientation, with the C-terminal of one facing the N-terminal of the other. Then the catalytic site has to form the amide bond by lowering the energy of activation.

$$HOOC - \overset{\overset{\displaystyle H}{|}}{\underset{\underset{\displaystyle R_1}{|}}{C}} - NH_2 \quad + \quad HOOC - \overset{\overset{\displaystyle H}{|}}{\underset{\underset{\displaystyle R_2}{|}}{C}} - NH_2 \quad + \quad HOOC - \overset{\overset{\displaystyle H}{|}}{\underset{\underset{\displaystyle R_3}{|}}{C}} - NH_2$$

C-terminals   N-terminal.   C- terminal   N-terminal.   C-terminal   N-terminal.

The above mentioned process can go on until it has formed a very long protein molecule. Once having produced the molecule, the enzyme will release it. It then passes through the rough endoplasmic reticulum until it can be packaged and moved to a vacuole and then to the cell membrane.

So now we know how cells do the following:

- Produce lipids and proteins

  Consume nutrients

  Eliminate wastes

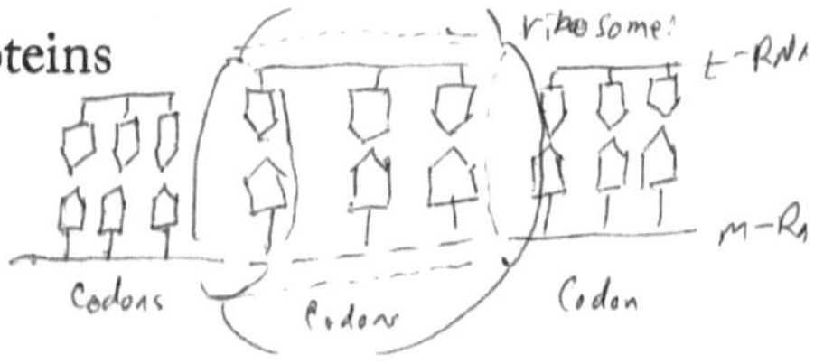

Before we leave this topic, let us look at the ribosome more carefully. It is a bead. It has a hole running through its center. As the messenger RNA moves through the center of the ribosome, it is read by transfer RNA, and a string of amino acids form.

Defining Life:

Let us take a minute to redefine life:

> Consumes nutrition
> Responds to stimuli
> Can reproduce
> Can grow
> Can adapt
> Can evolve

It is made of molecules that contain carbon. DNA and RNA make the above possible.

Let us talk further about protein synthesis.

First the mRNA and tRNA have to line up the amino acids that will be used to form the protein. Then the positioning site of the protease has to orient them so that the C-terminal of one faces the N-terminal of another. Then the catalytic site has to form the amide bond.

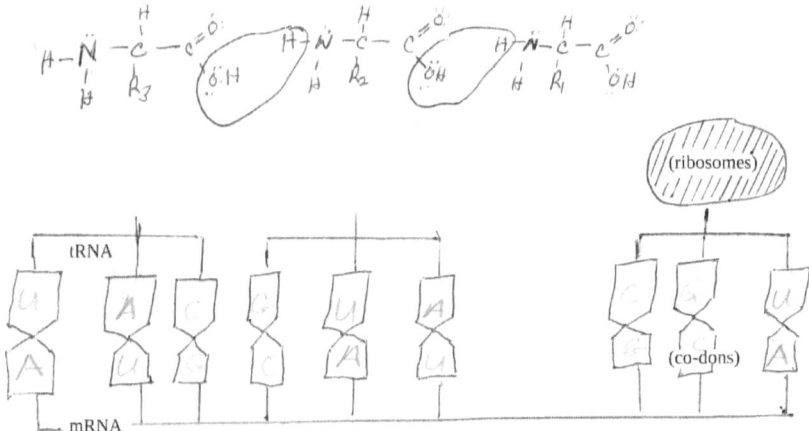

Triplicate code sequencing:

Protein molecule:

This is a theoretical arrangement.

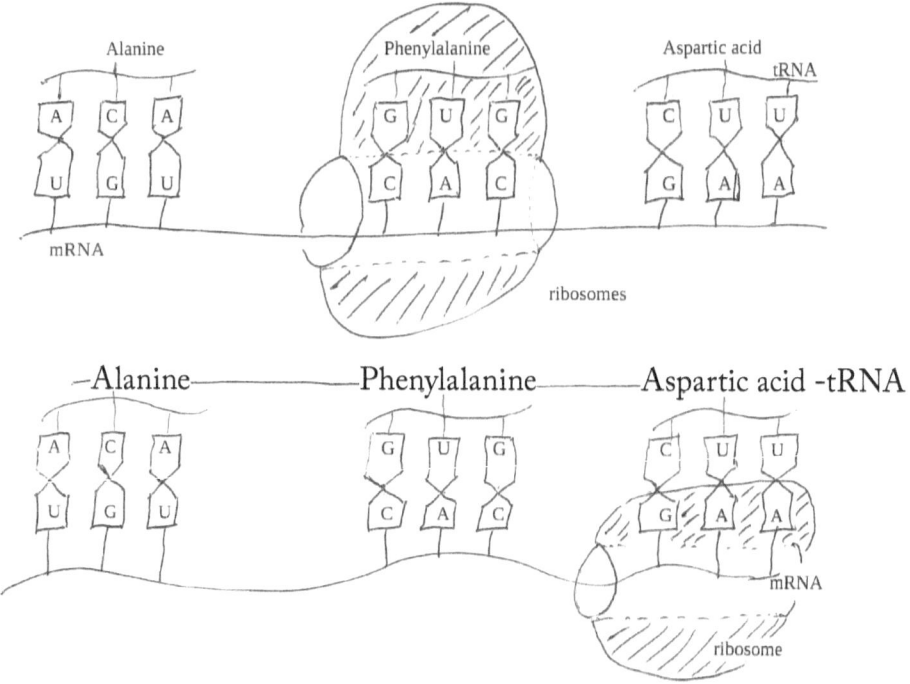

The protein is released and leaves the RER.

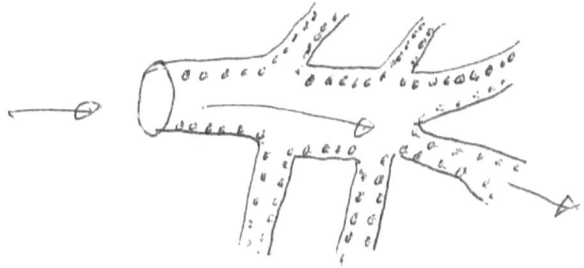

Let us consider the effect that mutations would have on protein synthesis. If a base were deleted or added, it would change the entire genetic code. It would create a "frame shift" mutation. This happens all the time and would be devastating if not corrected.

Let us see what happens when we have an addition mutation (which is rare).

Original:

55

Let us see what happens when we have an addition mutation (which is rare).

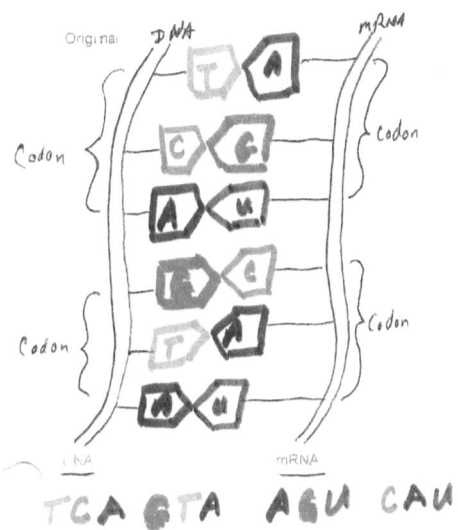

TCA GTA AGU CAU

Current sequence

Now let us add an additional base.

(another base has been added)

56

Now let us add an additional base

(another base has been added)

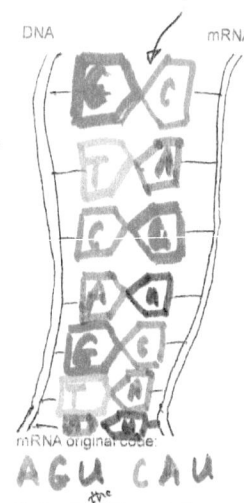

DNA   mRNA

mRNA original code:

AGU CAU

Code with the addition mutation:

CAG UCA

The new code is entirely wrong and the immune system must remove it.

Deletion Mutation

Original code: AGC UGC
New code: GCU GC

The new code is entirely wrong and unusable. It is up to the immune system to remove it.

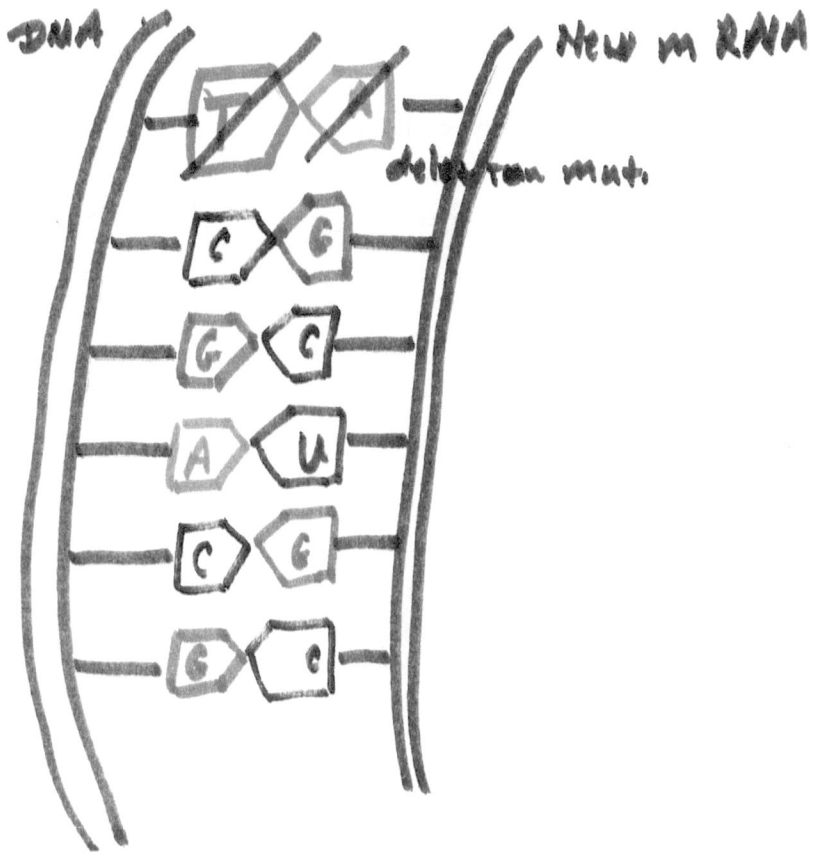

Deletion mutations are more common than addition mutations, but either would be devastating and both could be caused by:

- Viruses that add and remove DNA from the cells that they infect

  Toxins such as nitrites from spoiled foods

  Ionizing radiation such as alpha particles (α_4^2), which come from radon, a gas that leaks out of the Earth, beta rays (e)$^-$) from many radioactive elements, and X-rays, which come from the sun

  Heat, as with undescended testicles

  Natural aging

Review

The nucleosomes (made of histones) allow certain segments of DNA to open. The sense strand will then be free to code for a particular function. And when the gene is no longer needed the nucleosomes will allow it to close.

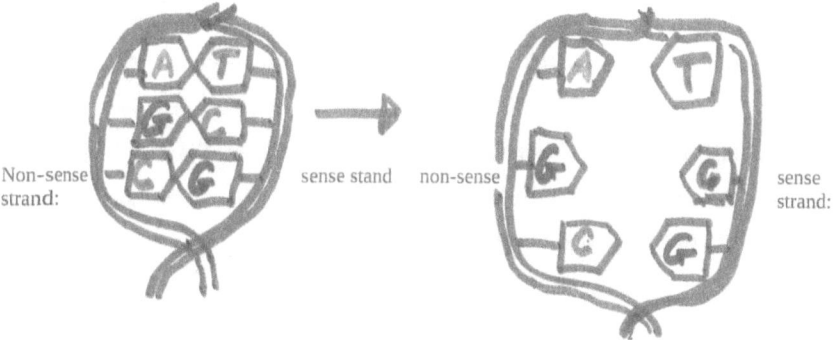

Non-sense strand:　　　sense stand　non-sense　　　sense strand:

The genetic code for DNA is divided into sets of three called codons. Every three nitrogen bases forms one codon unit, which codes for one amino acid.

Codon:
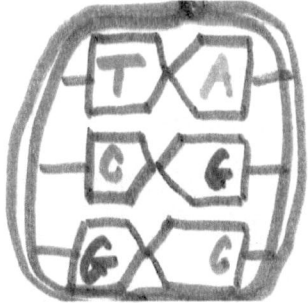

If anything were to cause a base to be added or deleted it would create a frame shift mutation. Let us say that the correct sequence is:

TAC GTA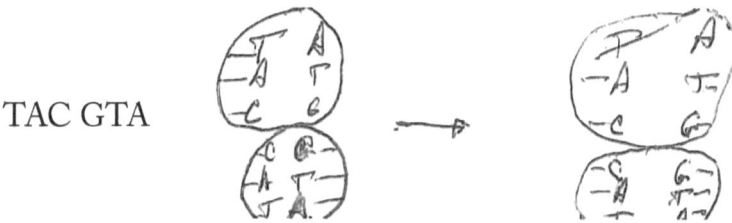

Supposing that something were to remove the first 'T'. This would be a deletion mutation and would lead to a frame shift phenomenon. The code would be read as:

ACG TA

This would obviously lead to an erroneous read. If we add a base the same thing would happen. Supposing we took the original TAC GTA and added an A? We would have ATA CGTA.

As we said, deletion mutations happen all the time, and the immune system must remove them. If the "sense" strand develops a frame shift mutation, the immune system must remove the affected segment. The "antisense" strand would then code for a new "sense" strand.

A mutation might also create a cancer cell, and the immune system would remove it. Immunity protects us against genetic defects and cancer.

Cellular Respiration

Let us change topics again and discuss energy production. Energy is the potential to perform useful work. It is generated by an organelle called the mitochondria and comes in the form of ATP (adenosine triphosphate).

Within the mitochondria there are three systems: the Embden-Meyerhof-Parnas pathway, the Krebs cycles (also known as the Citrus cycles), and the ETS (electron transport system).

There are two kinds of respiration: anaerobic and aerobic. Anaerobic does not require oxygen and only generates two molecules of ATP for each molecule of glucose. It also forms fermentation products such as lactic acid (2-hydroxy propionic acid), which flavors cheese and yogurt, and ethanol, which is the basis of alcoholic beverages. Lactic acid causes muscle cramps and sour milk and is metabolized by the liver.

Lactic acid (2-hydroxy propionic acid

$$CH_3-\underset{\underset{H}{|}}{\overset{\overset{\ddot{O}H}{|}}{C}}H-COOH$$

Ethanol:

$$CH_3-CH_2-\overset{O}{\underset{H}{C}}$$

Single-celled organisms usually use anaerobic respiration, though some will use aerobic. When sewers are constructed, they allow air to enter so that aerobic bacteria can flourish. They break down organic molecules more completely so there will be less smell and less explosive gas (such as methane).

Methane:

$$H - C - H$$

(with H above and H below the central C)

Multicellular life forms are more likely to use aerobic respiration because it provides more energy. It takes energy to create and maintain a multicellular structure. One reason is that multicellular forms require a form of intercellular "glue." Synthesizing it requires energy.

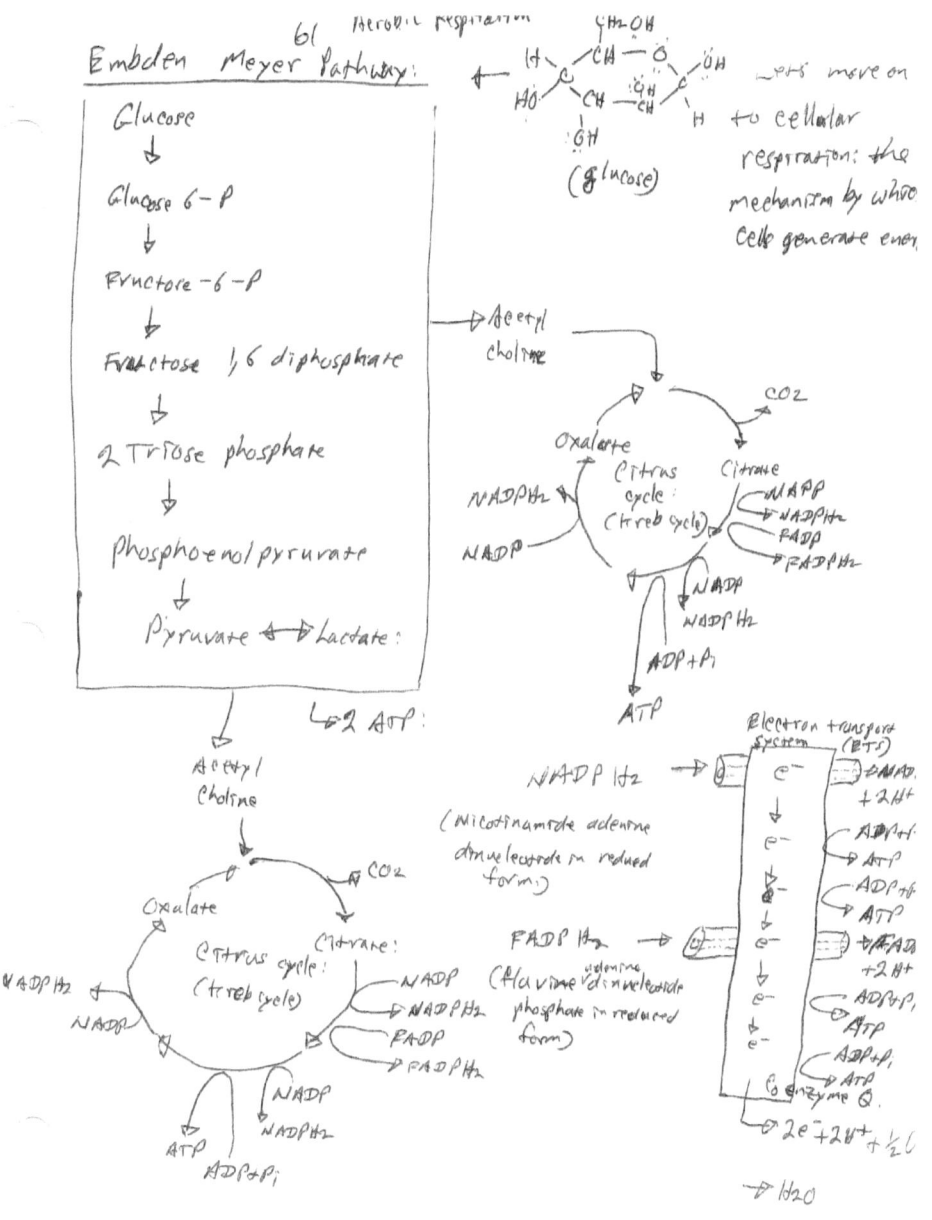

Aerobic respiration

Embden Meyer Pathway:

Glucose
↓
Glucose 6 - P
↓
Fructose - 6 - P
↓
Fructose 1,6 diphosphate
↓
2 Triose phosphate
↓
Phosphoenol pyruvate
↓
Pyruvate ↔ Lactate:

...ers move on to cellular respiration: the mechanism by which cells generate ener...

(glucose)

→ Acetyl choline

↳ 2 ATP:

Acetyl Choline

Oxalate

Citrus cycle (treb cycle)

Citrate

NADPH2
NADP
NADP
NADPH2

ATP
ADP+Pi

CO2

NAPP
NADPH2
FADP
FADPH2

NADP H2 → (Nicotinamide adenine dinucleotide in reduced form)

FADP H2 → (Flavine adenine dinucleotide phosphate in reduced form)

Electron transport system (ETS)

e⁻
↓
e⁻
↓
e⁻
↓
e⁻
↓
e⁻
↓
e⁻

NAD +2H⁺
ADP+ → ATP
ADP+ → ATP
FAD +2H⁺
ADP+Pi → ATP
ADP+Pi → ATP
Co enzyme Q

→ 2e⁻ + 2H⁺ + ½O
→ H2O

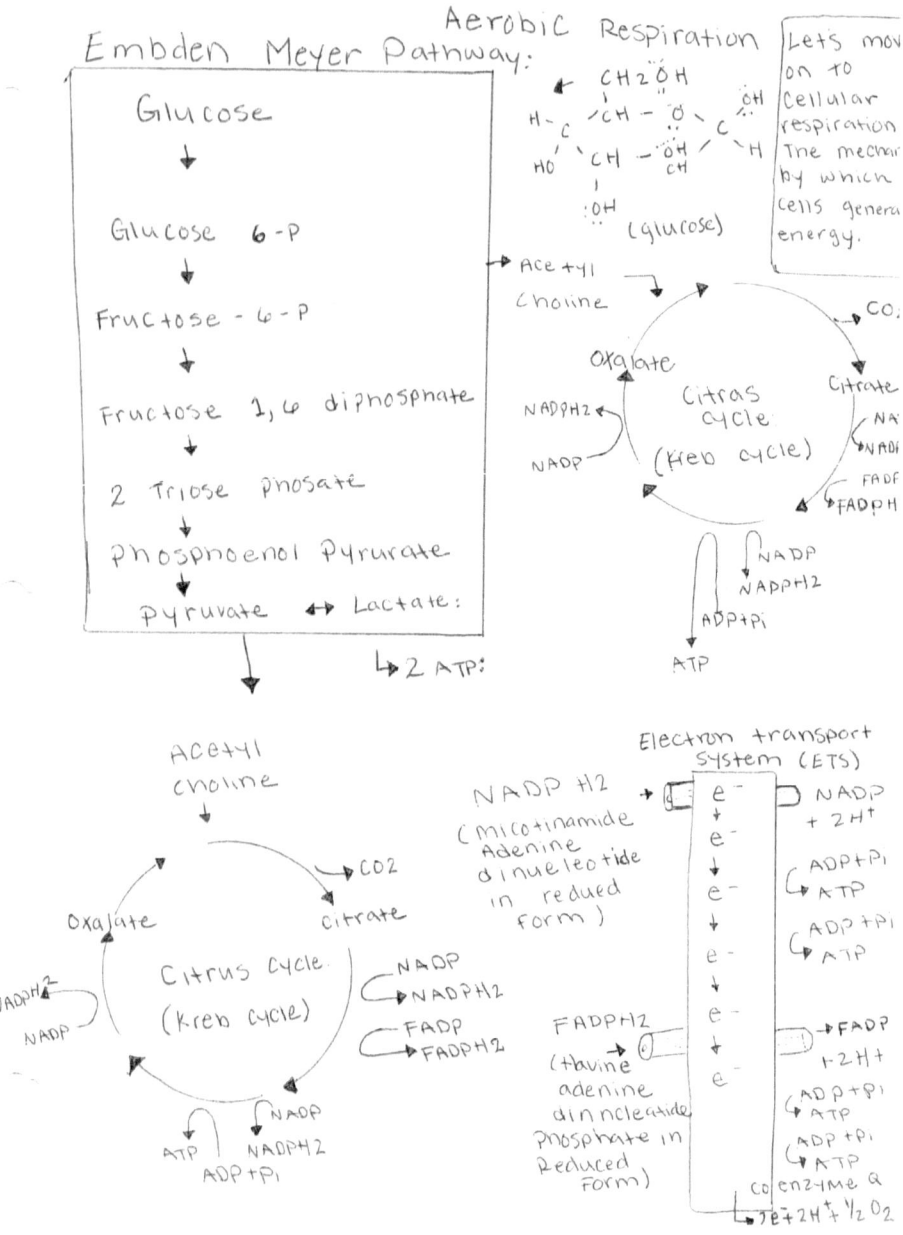

Embden Meyer Pathway:

Aerobic Respiration

Glucose

Glucose 6-P

Fructose - 6-P

Fructose 1,6 diphosphate

2 Triose Phosate

Phosphoenol Pyrurate

Pyruvate ↔ Lactate:

↳ 2 ATP:

CH2OH
H-C   /CH — O
HO CH -OH/   C—H
        CH
     :OH
     (glucose)

Acetyl Choline

Lets move on to Cellular respiration The mechan by which cells genera energy.

Oxalate

Citras cycle (Kreb cycle)

NADPH2

NADP

Citrate

CO₂

NA
NADF

FADF
FADPH

NADP
NADPH2

ADP+Pi

ATP

Acetyl choline

Oxalate

Citrus cycle (Kreb cycle)

NADPH2
NADP

CO2

citrate

NADP
NADPH2

FADP
FADPH2

NADP
NADPH2

ATP
ADP+Pi

Electron transport System (ETS)

NADP H2
(micotinamide Adenine dinucleotide in redued Form)

FADPH2
(flavine adenine dinucleatide Phosphate in Reduced Form)

e⁻
e⁻
e⁻
e⁻
e⁻
e⁻
e⁻
e⁻

NADP + 2H⁺

ADP+Pi
ATP

ADP+Pi
ATP

FADP +2H+

ADP+Pi
ATP

ADP+Pi
ATP

coenzyme Q
↳ 2e⁻+2H⁺ ½O₂

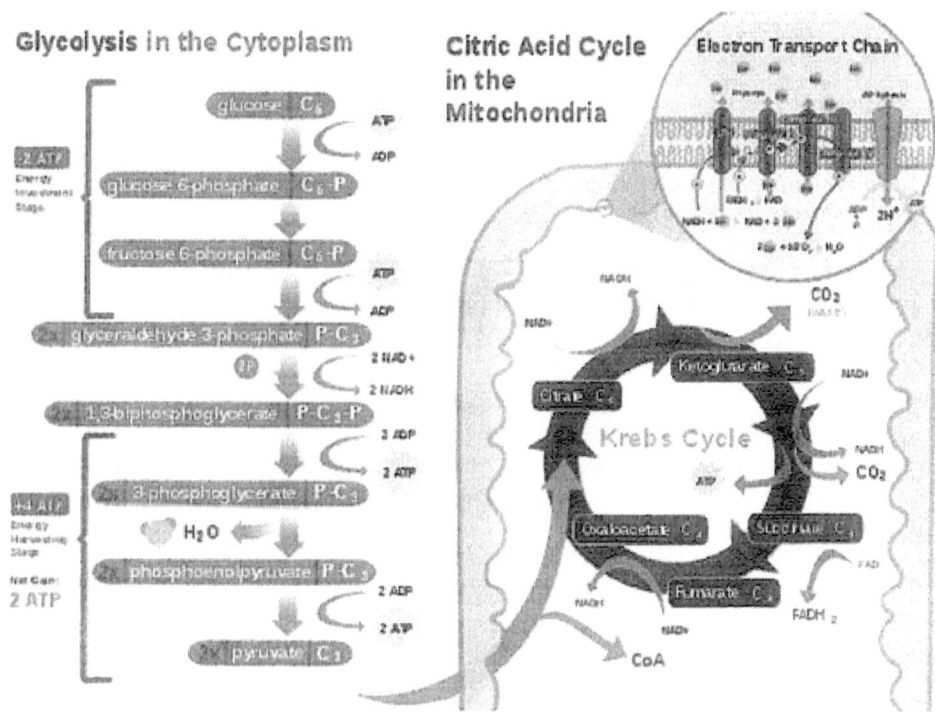

Glycolysis in the Cytoplasm

Citric Acid Cycle in the Mitochondria

Anaerobic respiration (without oxygen) and aerobic respiration (with oxygen) both take place in the mitochondria. It starts with a sugar called glucose. With anaerobic respiration, it is only metabolized by the Embden Meyerhof (EM) pathway and leads to the formation of two molecules of ATP. With aerobic, it passes from the Embden Meyerhof (EM), to the Kreb cycles, and to the Electron Transport System (ETS), and yields 36 molecules of ATP.

In the final step, electrons (e-), hydrogen ions (H+), and oxygen combine under the influence of coenzyme Q to form water.

$$2H^+ + 2e^- + \tfrac{1}{2}O_2 \text{(air)} \xrightarrow{\ C_oQ\ } H \overset{\cdot\cdot}{\underset{H}{O}} \text{(water)}$$

If we lost our ability to breathe, we would die of "acidosis." The concentration of hydrogen ions would be so high (and pH would be so low) that our enzymes would cease to function.

Mitochondria have their own DNA and divide separately. When cells get ready to reproduce, they somehow trigger the mitochondria into dividing as well. Some biologists believe that mitochondria were once free-living organisms that set up a symbiotic relationship with higher cells (a relationship that benefits both).

Biosynthesis

Now let us switch gears once again and talk about the production of different molecules. To understand this, we must first realize that molecules contain positive and negative electric charges known as "moments" (see diagram). They also form Van der Waal bonds, and we have to know that electrons will spend more time orbiting an atom that has a high atomic number (many protons) and less time with atoms that have small atomic numbers (few protons). If electrons spend more time orbiting one atom, it will have a slight negative charge (moment$^-$). If they spend less time, it will have a positive charge (moment$^+$).

Let us talk about how the SER makes lipids (i.e., triglycerides):

hydroxyl groups:
$$CH_2 - CH - CH_2 \quad glycerine$$
$$O: \quad :O: \quad :O:$$
$$H \qquad H \qquad H$$

carboxyl groups:
$$:O \quad :OH \quad :O \quad :OH \quad :O \quad :OH$$
$$C \qquad C \qquad C \qquad fatty\ acid$$
$$(CH_2)_{n_1} \quad (CH_2)_{n_2} \quad (CH_2)_{n_3}$$
$$CH_3 \qquad CH_3 \qquad CH_3$$

positioning site:

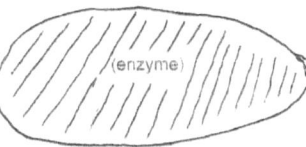

(enzyme)

glycerine

hydroxyl groups

carboxyl groups

fatty acids

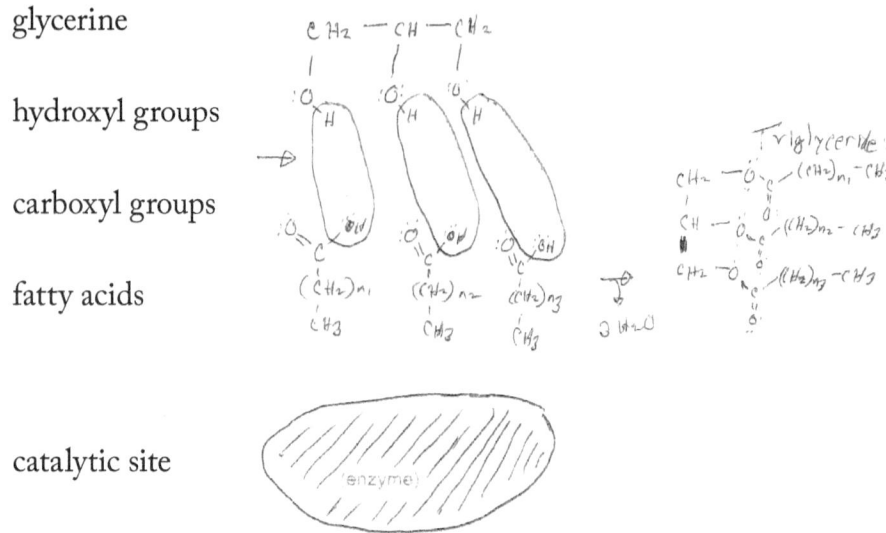

catalytic site

Formation of Triglyceride

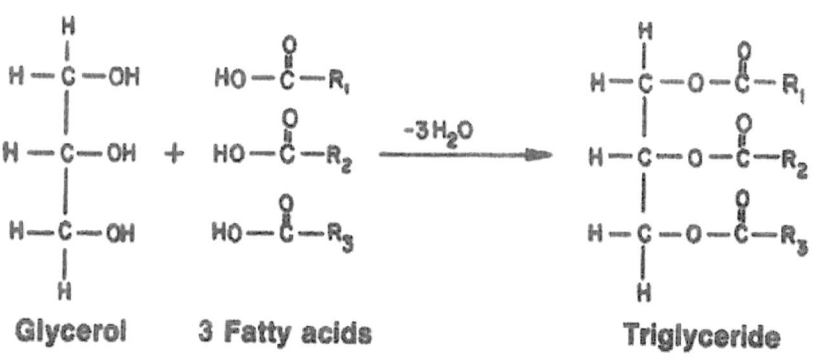

Before we leave the topic of synthesis let us expand upon one final point. How does the Smooth Endoplasmic Reticulum (SER) synthesize lipids? Let us look at two lipids in particular:

Triglycerides are formed when one molecule of glycerin combines with three molecules of fatty acid. To catalyze this, we need an enzyme called triglyceride synthetase, which works in a way similar to protein polymerase. But instead of forming amide bonds, it forms ester bonds—bonds between a hydroxyl group and a carboxyl group.

Hydroxyl: $-CH_2-\ddot{O} \diagdown H$ Carboxyl: $-CH_2-C \diagup\!\!\!\!\!\diagdown \!\!\!\!^{O}_{\ddot{O}-H}$

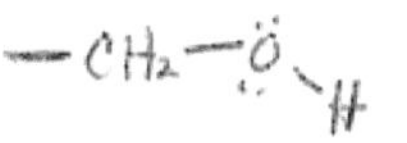

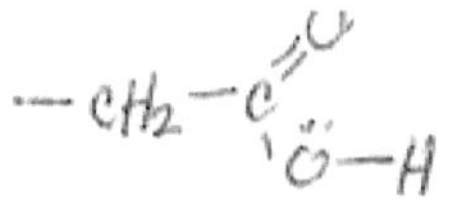

Again there will be a dehydration reaction that will generate three molecules of water, and an esterification reaction in which ester bonds will form. This will be carried out by triglyceride synthetase, which will have a positioning site and a catalytic site.

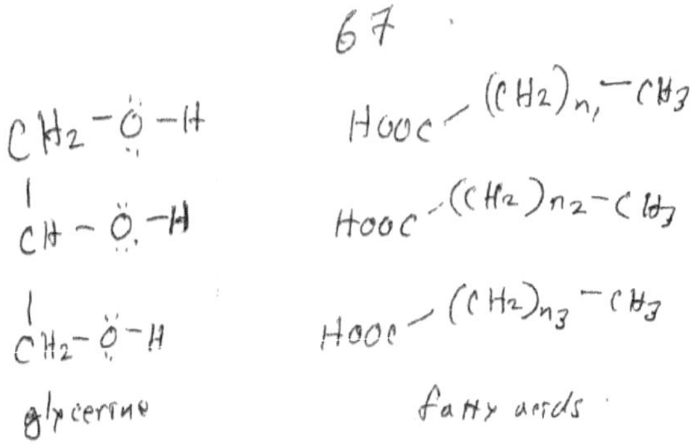

67

$CH_2-\ddot{O}-H$

$CH-\ddot{O}-H$

$CH_2-\ddot{O}-H$

glycerine

$HOOC-(CH_2)_{n_1}-CH_3$

$HOOC-(CH_2)_{n_2}-CH_3$

$HOOC-(CH_2)_{n_3}-CH_3$

fatty acids

And the enzyme's positioning site has to align all four molecules correctly.

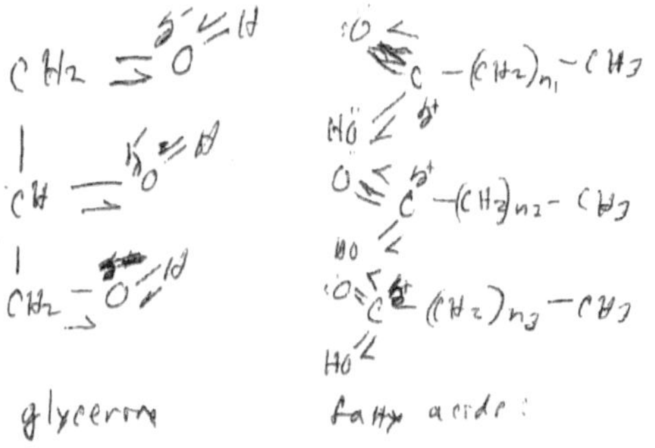

$CH_2 = O = H$

$CH = O = H$

$CH_2 - O$

glycerin

$O = C - (CH_2)_{n_1} - CH_3$
HO

$O = C - (CH_2)_{n_2} - CH_3$
HO

$O = C - (CH_2)_{n_3} - CH_3$
HO

fatty acids :

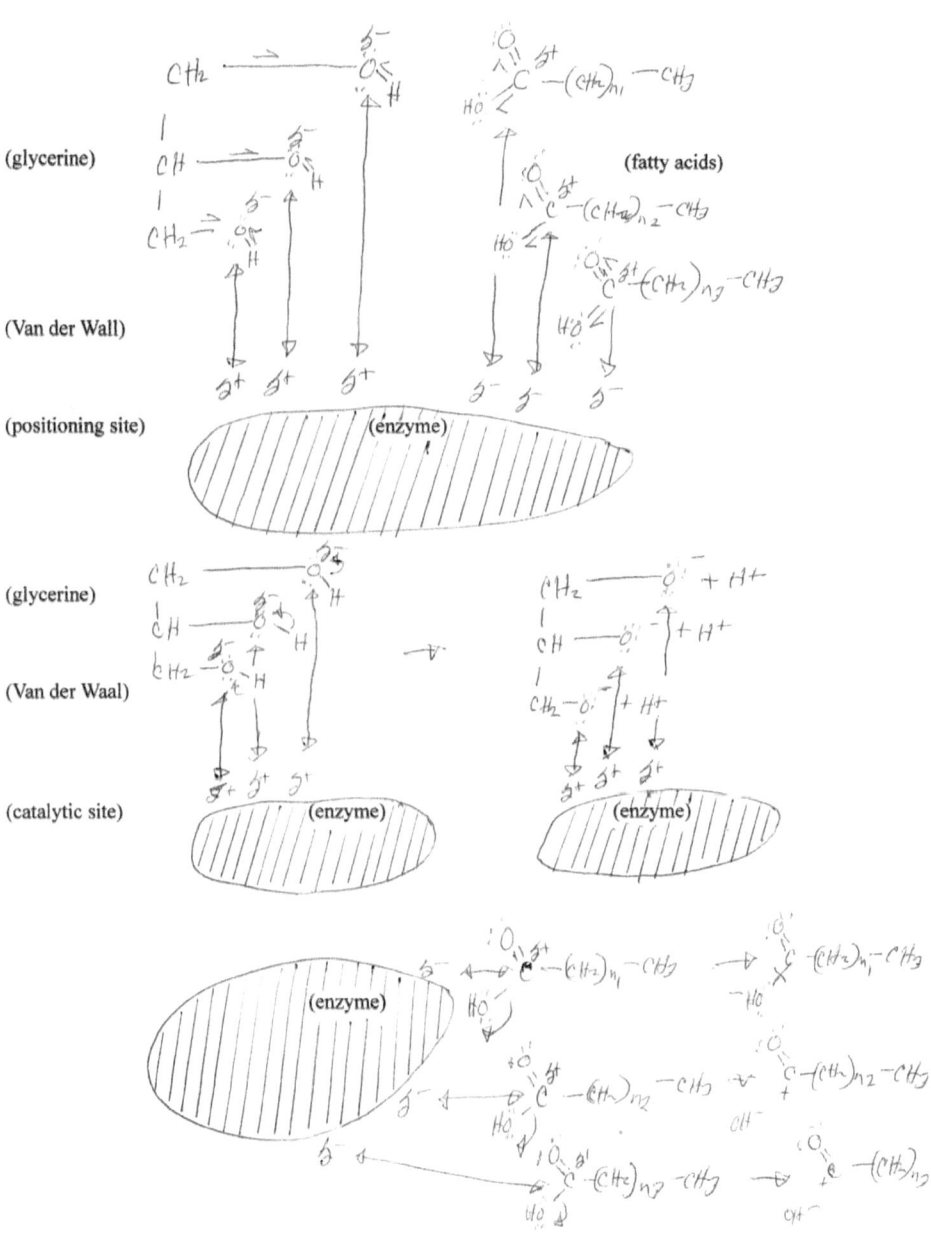

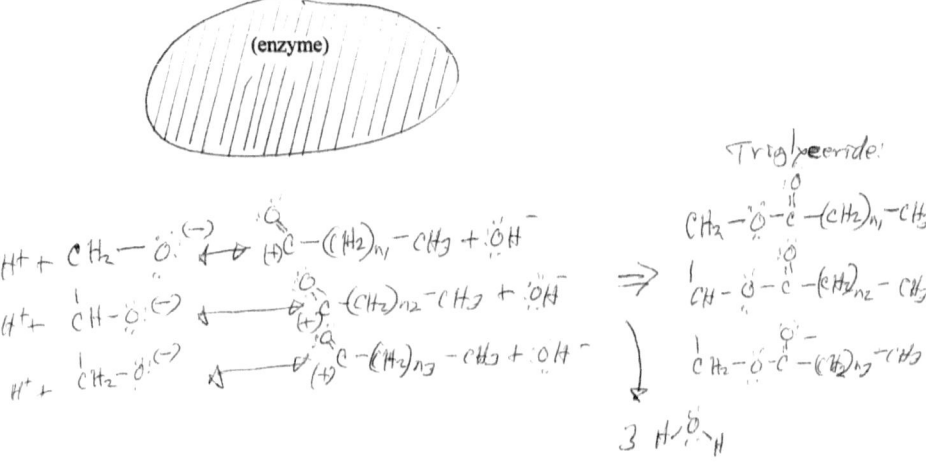

There is one more topic that we should discuss before leaving chemistry: the composition, structure, and activity of enzymes. Enzymes are biological catalysts. They promote chemical reactions in biological systems without being consumed by them. Enzymes are proteins which contain cofactors, such as trace metals, and coenzymes (usually synthesized from vitamins). When systems create enzymes, they link amino acids. In doing so, they form a long protein polymer. This molecule will twist upon itself to form an alpha helix, a springlike structure. The alpha helix will then twist upon itself again:

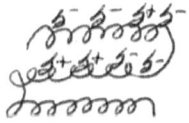 (α - helix with partial positive and negative charges called moments:)

In the end we get a long protein molecule that is folded upon itself many times. There will be a distribution of partial charges upon this molecule that are referred to as "moments":

The substrate that the enzyme interacts with also contains positive and negative moments. The moments interact with each other through electrostatic attraction. The force that acts between them is referred to as a Van der Waal bond, and it is this mechanism that guides two substrate molecules into the correct orientation: in example, the N-terminal of one amino acid faces the C-terminal of another. This is the enzyme's position site:

Substrate
Van der Waal bonds.

enzyme.

Addendum
(the feedback mechanism of enzymes)

During our discussion of enzymes we said that they have a positioning site that aligns the substrate molecules, and a catalytic site that forms the bond. Then we said that a "feedback" mechanism tells the enzyme to let go of the final product. What exactly is this "feedback" mechanism?

To answer this, let us look at protein synthesis once again.

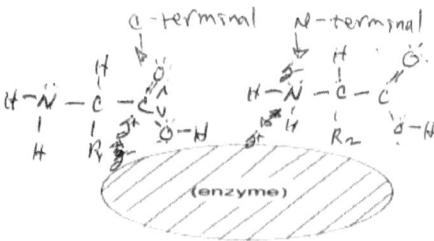

First the position site aligns two molecules of amino acid so that the N-terminal of one faces the C-terminal of another:

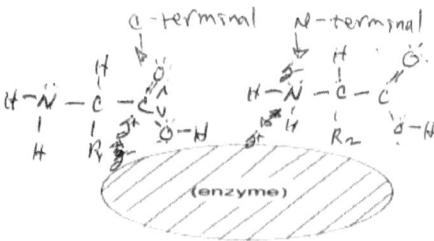

Then an amide bond forms. This is the work of the catalytic site.

The product is now at its lowest energy state.
But two bonds have to go.

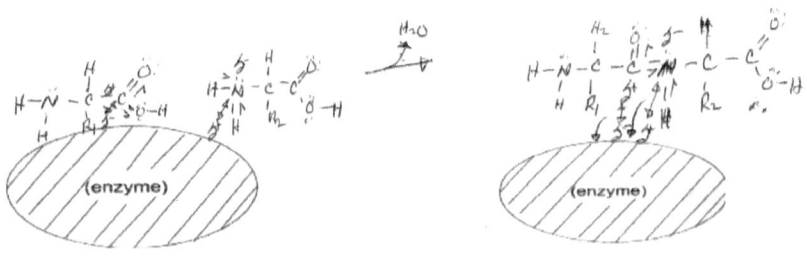

Now we have a two–amino acid protein that needs to be released.
What tells the enzyme to release it? Let us look at it more carefully.

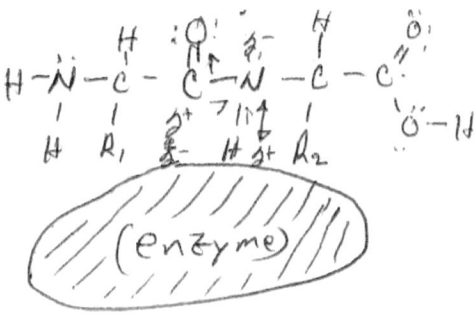

It is still attached to the enzyme by Van der Waal bonds, and there
are five bonds going to the carbon and four to the nitrogen. But
carbon is tetravalent (only supposed to have four) and nitrogen is
trivalent (only supposed to have three).

Something has to be rearranged. But what are we going to
rearrange? The bonds within the protein molecule have already
reached their lowest point of enthalpy (the lowest possible energy
state). So it would not make thermodynamic sense for them to
rearrange further. The only alternative would be for the enzyme to
withdraw its Van der Waal bonds, and in doing so it satisfies two

requirements. It creates the correct number of bonds and allows the finished product to be released.

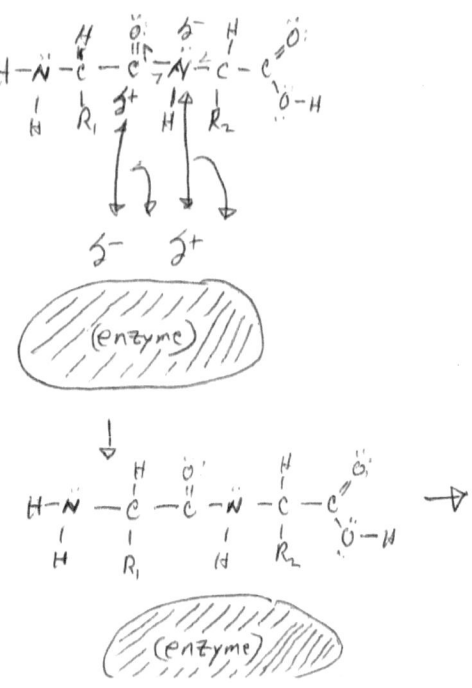

The protein molecule is at the lowest energy state. But the carbon still has five (5) bonds and the nitrogen has four (4). Two of the bonds have to go, and it only makes sense that it should be the Van der Waal bonds.

Once created a feedback mechanism releases the final product.

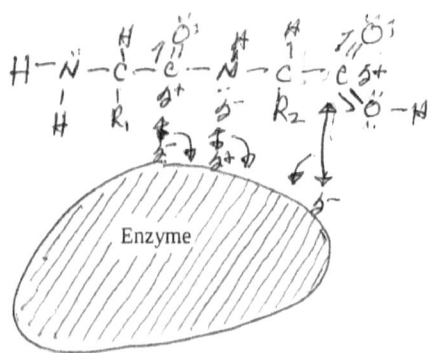

Feedback releases the protein from the enzyme.

Synthesis of Glycoproteins

How do cells synthesize glycoproteins?

They need:

1. One molecule of glucose
2. One molecule of protein

They will create a molecule with a positively charged carbon and a negatively charged nitrogen. They will also synthesize one molecule of water (H_2O). It will be similar to triglyceride synthesis.

Glucose:

Protein:

Glycoprotein:

Johannes Van der Waal
1873

Net dipole

$\mu = 1.47D$

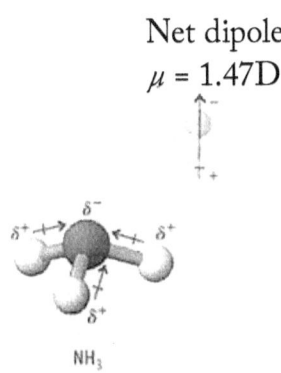

NH_3

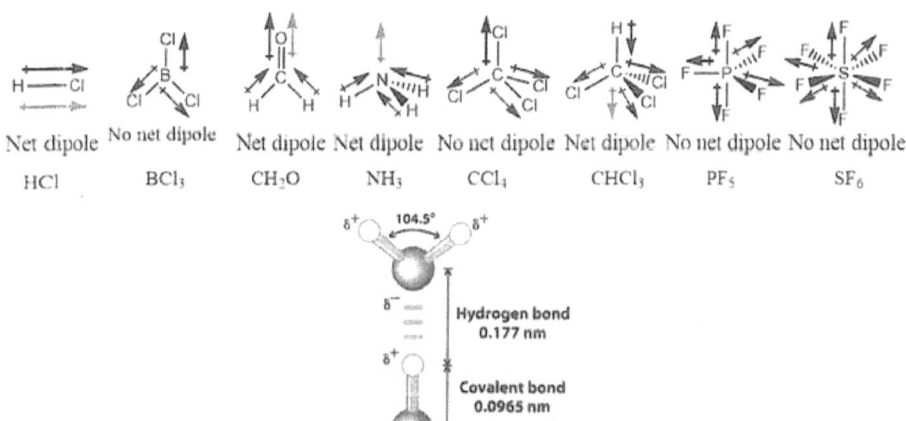

Net dipole	No net dipole	Net dipole	Net dipole	No net dipole	Net dipole	No net dipole	No net dipole
HCl	BCl_3	CH_2O	NH_3	CCl_4	$CHCl_3$	PF_5	SF_6

Van der Waal's Forces (VDW)
Diagram

KEY

\+ POSITIVE NUCLEUS

— NEGATIVE CHARGED ELECTRON CLOUD

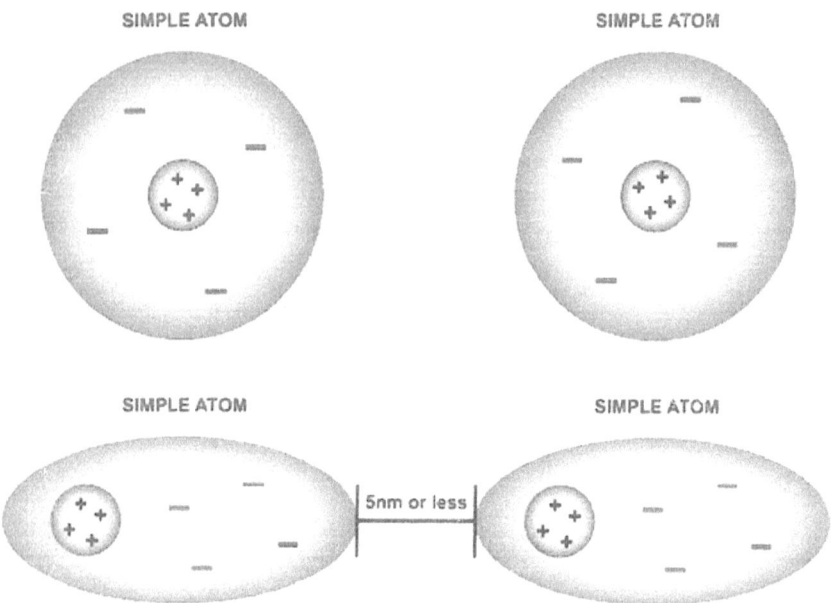

When two atoms come within 5 nanometers of each other, there will be a slight interaction between them, thus causing polarity and a slight attraction.

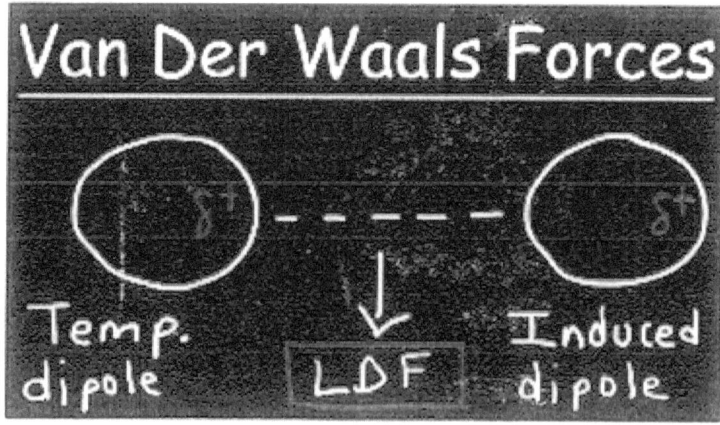

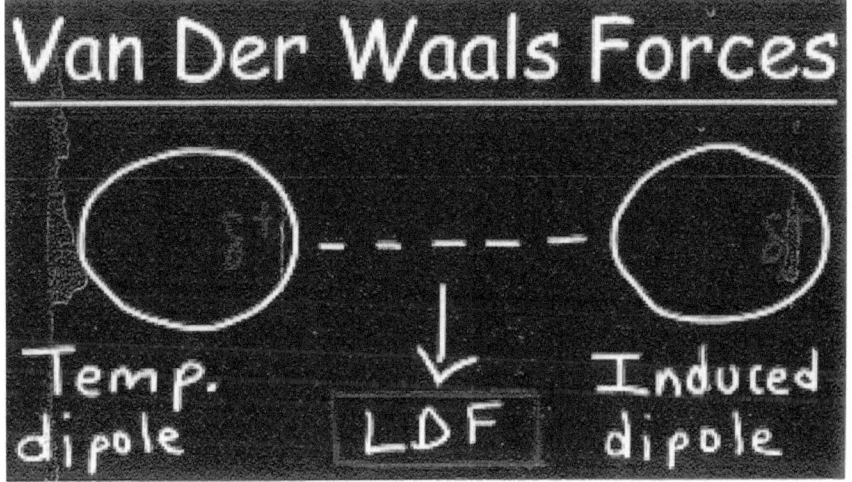

Ester Formation

carboxyl   hydroxide

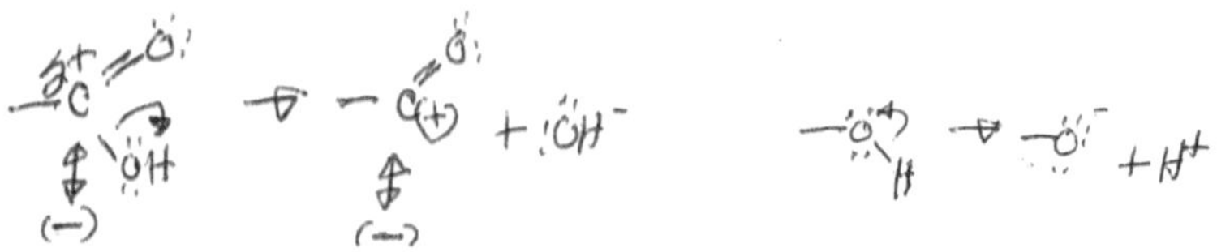

The H^+ and OH^- combine to form water:

H+   +   OH-   ⟶   H—O—H

And the negatively charged oxygen is attracted to the positively charged carbon.

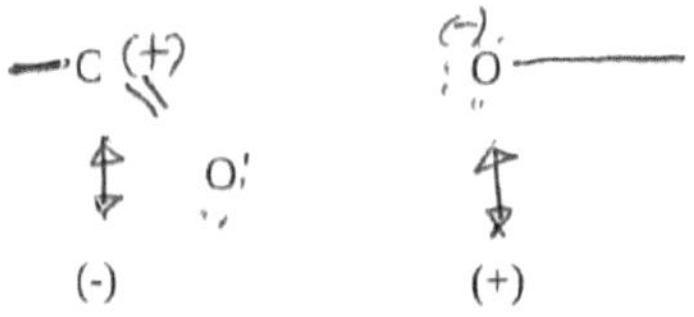

—C (+)   (-) O ————

(-)   (+)

The ester is formed, and a feedback mechanism tells the enzyme to release.

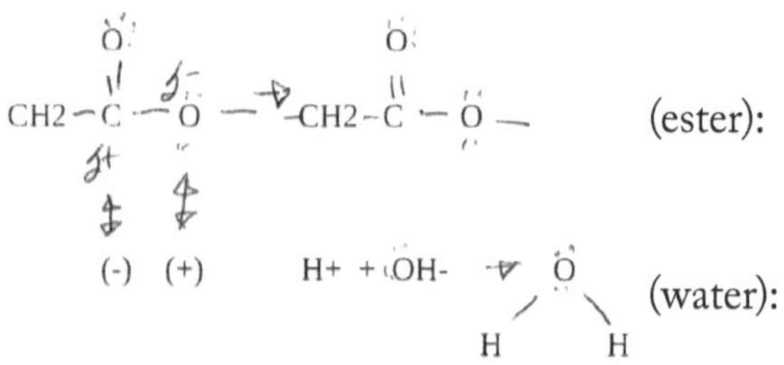

CH2–C—O ⟶ –CH2–C – O —   (ester):

(-) (+)   H+ + OH- ⟶ O   (water):
                        H   H

Let us talk about the synthesis of lipoproteins.

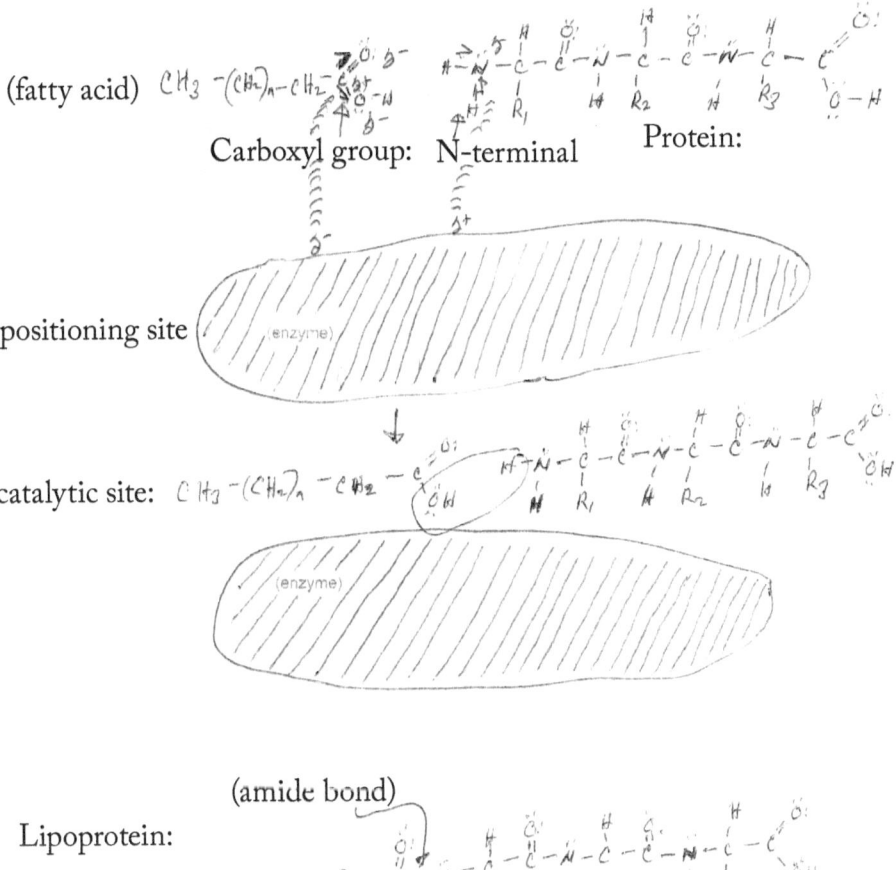

(fatty acid)

Carboxyl group: N-terminal   Protein:

positioning site

catalytic site:

(amide bond)

Lipoprotein:

Protein Synthesis

But again the nitrogen and the carbon have too many bonds. The molecule has to rearrange.

The enzyme withdraws the Van der Waal bonds.

Final product:

Review:

The positioning site of the enzyme has to align the molecule correctly.

Let us take a minute to review the seven characteristics of life:

1. Consumes nutrition
2. Responds to stimulus
3. Reproduces
4. Evolves
5. Adapts
6. Grows
7. Made of molecules that contain carbon

We now understand how cells:

1. Consume nutrition:
   - Pores in the cell membrane
     Pinocytosis
     Phagocytosis

2. Grow: they synthesize lipids and proteins.

3. They consist of molecules that contain carbon.

The other characteristics will become clear after we discuss genetics.

Let us move on to a discussion of the organelles.

Golgi bodies:

Golgi bodies are vessels that contain lipids. Lipids are a group of compounds that:

- Are nonpolar: there is an even distribution of electric charge throughout the molecule:

  They dissolve in organic solvents and detergents, but not in water.

  They do not dissolve salts.

  They serve as:

    Insulation

    o   Parts of cell membranes

    Producers of energy

    In the synthesis of hormones, lipoproteins, and sphingolipids

They include:

- Oils

  Waxes
  Triglycerides
  Precursors of myelin sheaves
  Cholesterol

Cells synthesize lipids in the smooth endoplasmic reticulum (SER). So there are two ways for Golgi bodies to obtain them:

- Consume them from outside the cell by phagocytosis.

  Obtain them from the smooth endoplasmic reticulum.

Once having obtained lipids, Golgi bodies could deliver them to another part of the cell or take them to the cell membrane to be released by reverse phagocytosis.

Competitive and Non-Competitive Inhibition

When talking about the activity of enzymes, we need to define two terms:

    Competitive Inhibition
    Non-competitive Inhibition

As we said, enzymes have two sites, a positioning site and a catalytic site. The catalytic site will bind two molecules of reactant, which we will call A and B, and form a product that we will call C.

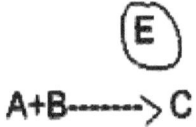

$$A + B \text{------} > C$$

And once C has been synthesized, the enzyme will release it. But suppose that we gave the enzyme a chemical that was similar to A and binds the catalytic site, but dissimilar enough not to react with B to form C? We will call this chemical A'.

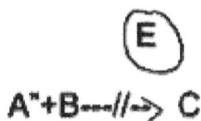

$$A' + B \text{---} // \text{-} > C$$

The enzyme would be neutralized. Its catalytic site would be blocked by A'. A' would adhere to the catalytic site, but it could not become C, so the enzyme cannot release. There are two ways to overcome competitive inhibition.

First would be to provide the enzyme with a high concentration of the correct substrate A. The second would be time. If you had time, the A' would disengage by itself.

Competitive inhibition is the mechanism for most medications such as antibiotics, sedatives, anesthetics, and drinking alcohol (ethanol).

Non-competitive inhibition involves binding enzymes at some site other than the catalytic site. It is the mechanism of most toxins and can sometimes be reversed with antidotes. Cyanide can be treated with thiosulfate. Carbon monoxide can be treated with oxygen, and methanol toxicity can be treated with ethanol.

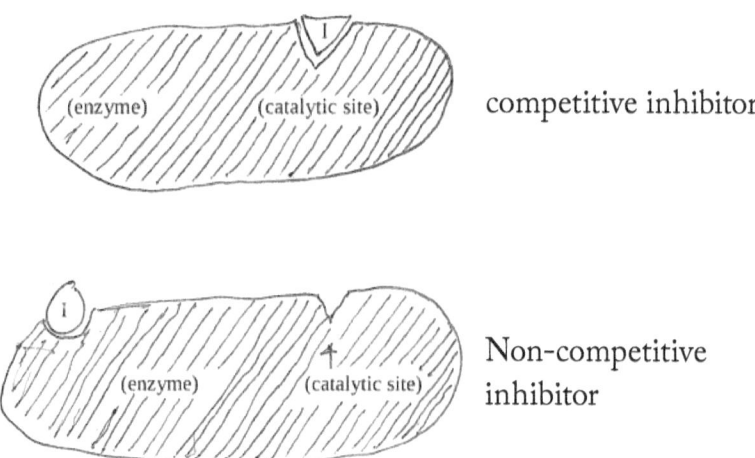

competitive inhibitor

Non-competitive inhibitor

Let us review the production of three lipids:

Triglycerides:

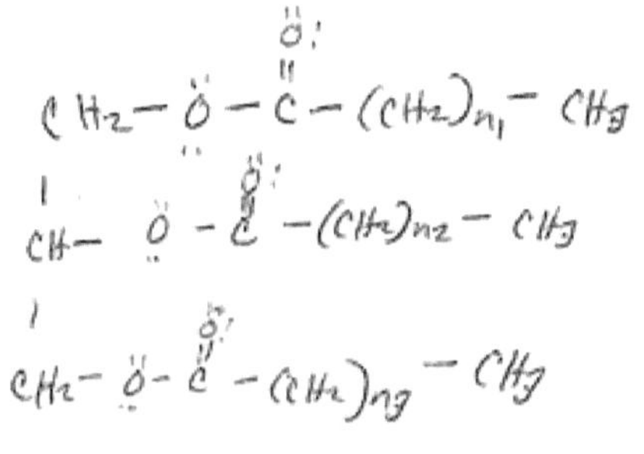

$$CH_2-O-\overset{O}{\overset{\|}{C}}-(CH_2)_{n_1}-CH_3$$
$$CH-O-\overset{O}{\overset{\|}{C}}-(CH_2)_{n_2}-CH_3$$
$$CH_2-O-\overset{O}{\overset{\|}{C}}-(CH_2)_{n_3}-CH_3$$

Lipoprotein:

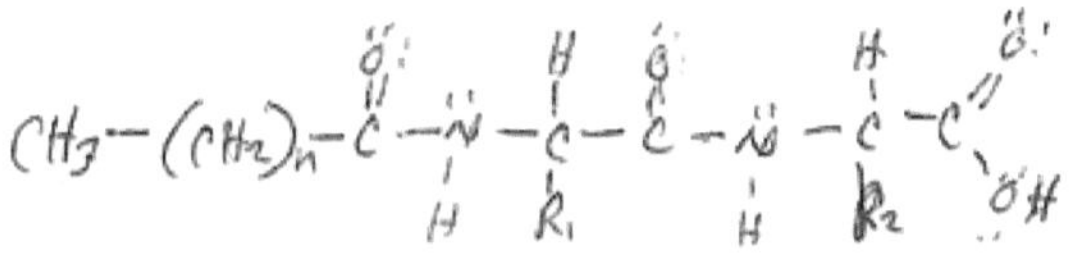

$$CH_3-(CH_2)_n-\overset{O}{\overset{\|}{C}}-\overset{H}{\underset{H}{N}}-\overset{}{\underset{R_1}{C}}-\overset{O}{\overset{\|}{C}}-\overset{H}{\underset{H}{N}}-\overset{}{\underset{R_2}{C}}-\overset{O}{\overset{\|}{C}}{\underset{OH}{}}$$

Glycolipid:

$$\begin{array}{c} CH_2OH \\ CH-O \\ H-C \\ HO-CH-CH \quad OH \\ OH \end{array} \quad O-\overset{O}{\overset{\|}{C}}-(CH_2)_n-CH_3$$

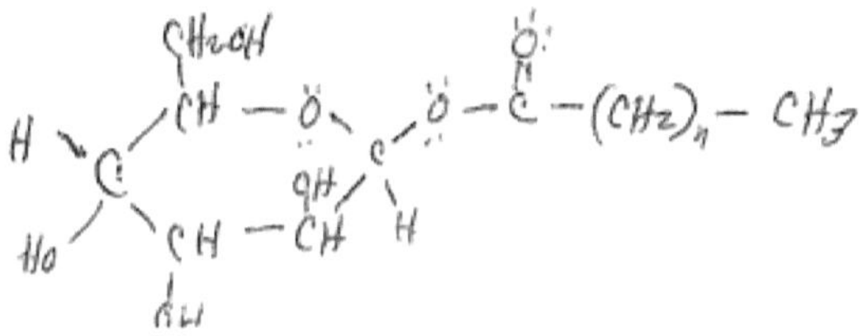

To synthesize triglyceride, we need:

- One molecule of glycerin

  Three molecules of fatty acid

Glycerine: Fatty acids: Triglycerides:

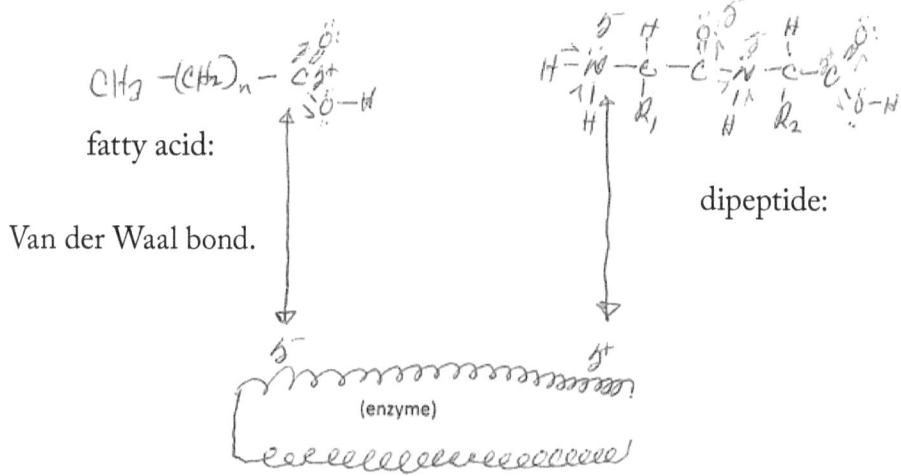

How do cells make lipoprotein? Let us begin with one molecule of a fatty acid and a dipeptide.

fatty acid:

Van der Waal bond.

dipeptide:

(enzyme)

The positioning site has to align the molecules so that the C-terminal of the fatty acid faces the N-terminal of the peptide. Then the catalytic site has to lower the energy of activation so that an amide bond can form.

With the Van der Waal bonds, there will be five bonds going to the carbon and four to the nitrogen—one too many in each case. The molecule has to rearrange.

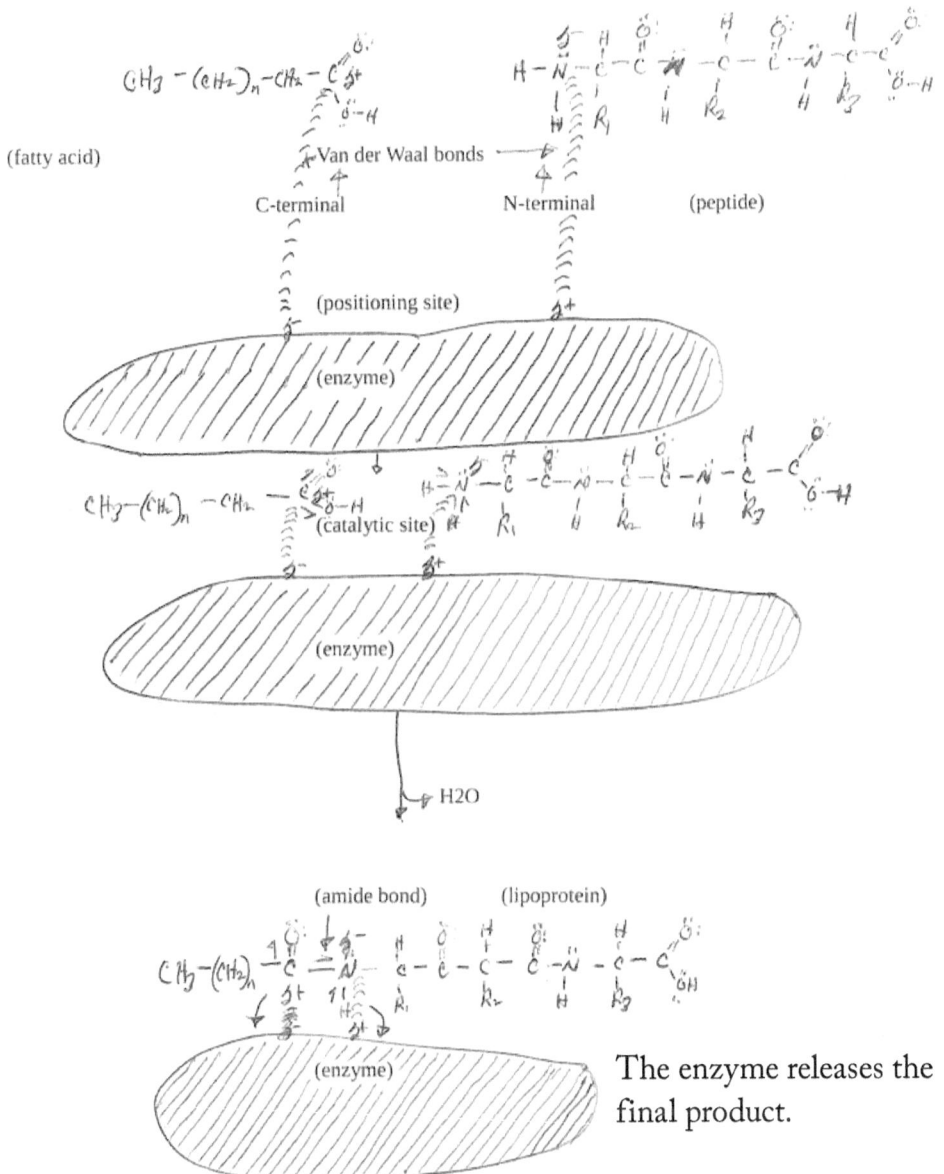

(fatty acid)

Van der Waal bonds

C-terminal   N-terminal   (peptide)

(positioning site)

(enzyme)

(catalytic site)

(enzyme)

H2O

(amide bond)   (lipoprotein)

(enzyme)

The enzyme releases the final product.

Catalytic Sites

Lowering energy
of activation when
forming an amide bond
or (an ester) bond.

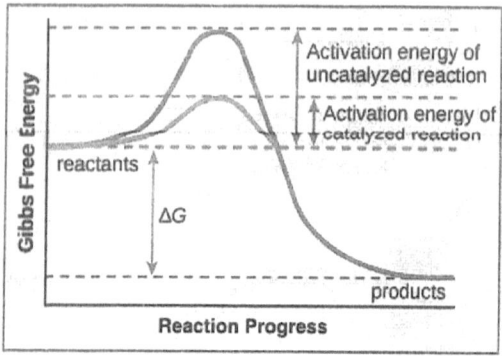

Image modified from "Potential, kinetic, free, and activation energy: Figure 5," by OpenStax College, Biology. CC BY 3.0.

To clarify one important point, enzymes don't change a reaction's ΔG value. That is, they don't change whether a reaction is energy-releasing or energy-absorbing overall. That's because enzymes don't affect the free energy of the reactants or products.

Instead, enzymes lower the energy of the **transition state**, an unstable state that products must pass through in order to become reactants. The transition state is at the top of the energy "hill" in the diagram above.

Active sites and substrate specificity

Enzyme-Substrate Interaction

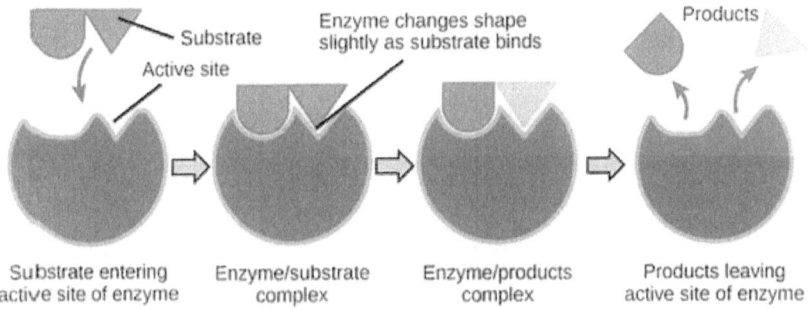

Substrate

Active site

Enzyme changes shape slightly as substrate binds

Products

Substrate entering active site of enzyme

Enzyme/substrate complex

Enzyme/products complex

Products leaving active site of enzyme

Two electrically charged sites have to come within 5 nanometers (nm) of each other for a Van der Waal bond to form: 5 billionths of a meter. So enzymes have to bring substrate molecules close together; i.e.,

Carboxyl
5 nanometers (nm) or less.
Enzyme

Let us move on to chemistry. It will greatly increase our understanding of biological systems. Let us begin by discussing:

- The atom

  Chemical bonding

  The composition of:

  Electrolytes

  o Lipids

  Amino acids

  Proteins

  Carbohydrates

  Nucleic acids

- Radicals

  Diffusion

  Osmosis

  Buffers

The Chemistry of Life

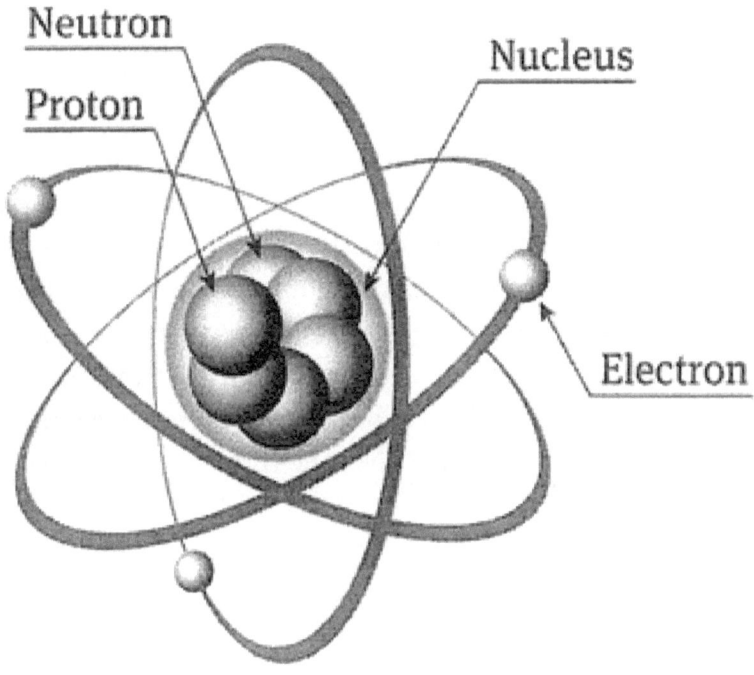

Atom

The atom consists of three particles: electrons, protons, and neutrons. Protons and neutrons make up the nucleus, and electrons orbit around it. This is also known as the Niels Bohr model, after the physicist who devised it in 1911. Occasionally it is referred to as the planetary model.

The nucleus is held together by strong and weak nuclear forces. The number of protons is referred to as the atomic number. The number of neutrons plus protons is the atomic weight. When talking about an element (a chemical that cannot be broken down further), one uses the symbol of the periodic table and places the atomic number to the right. In an example, helium has two protons and two neutrons, so the atomic number is two, and the atomic weight is four. The number of electrons is always equal to the number of protons. This is because electrons have a negative charge, and protons have a positive, and the charges must balance.

$$He_4^2$$

Atoms form bonds. There are two basic bonds:

- Covalent

Ionic

Ionic bonds form between two electrically charged atoms, known as ions. Atoms with a negative charge are anions. Those with a positive are cations. Negatives and positives attract. So anions attract cations, and when they come together, they form a bond.

Examples of ionic compounds (or chemicals made of two or more different elements) include:

Table salt Na^+ Cl^-

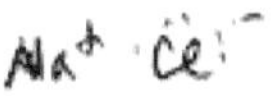

Baking soda

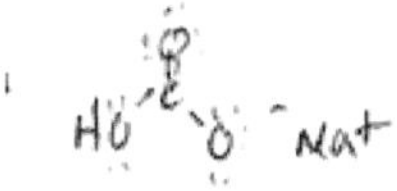

Liquid drain opener

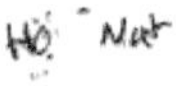

Soap

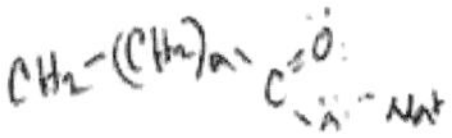

Covalent bonding involves the sharing of electrons. Examples include:

Water

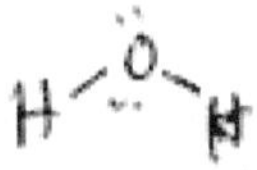

Sugar

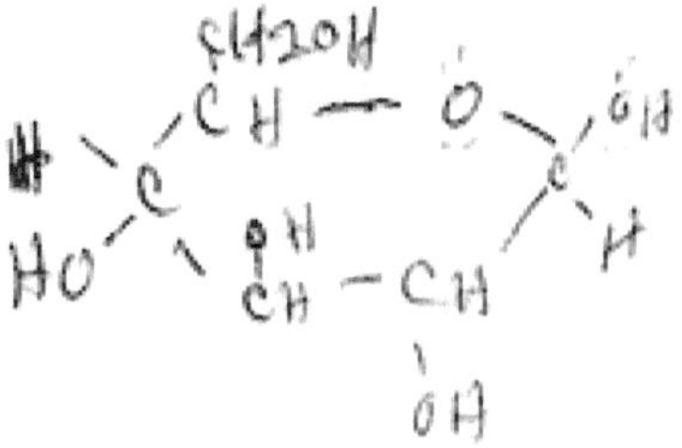

Natural gas

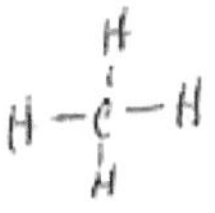

Drinking alcohol

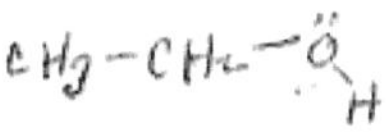

There are three kinds of covalent bonds:

Single bonds

Double bonds

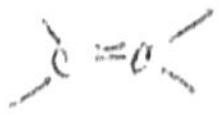

Triple bonds

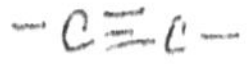

Triple bonds are unstable and are rarely found in nature. One exception is the cyanide side chain. But single and double bonds are common.

Let us move on to a discussion of different compounds:

Electrolytes: chemicals that allow water to conduct electricity. Pure water will not conduct.

But if you add electrolytes, it will.

In water, electrolytes form ions. Anions and cations are what allow an electric current to flow.

Electrolytes include:

Table salt $Na^+ \cdot Cl^-$

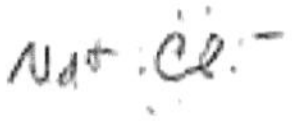

Epsom salt $Mg^{2+}\ ^-\ddot{O}-\overset{\overset{\ddot{O}}{\|}}{\underset{\ddot{O}}{S}}-\ddot{O}^-$

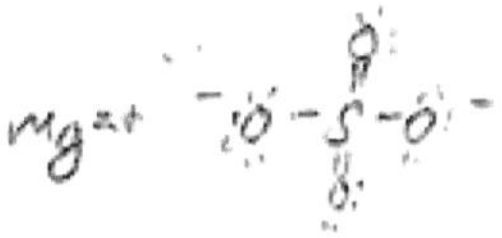

Bleach $\ddot{C}l-\ddot{O}^-\ Na^+$

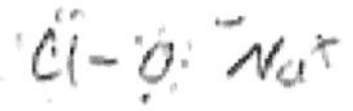

We need electrolytes to live.

Lipids do not dissolve in water but do dissolve in alcohols, detergents, and organic solvents. They include:

• Waxes

   Cholesterol
   Triglycerides
   Lipoproteins
   Phospholipid parts of cell membranes
   Fatty acids
   Oils

Biological Molecules

Let us talk about some of the molecules commonly found in biology.

The structure of amino acids:

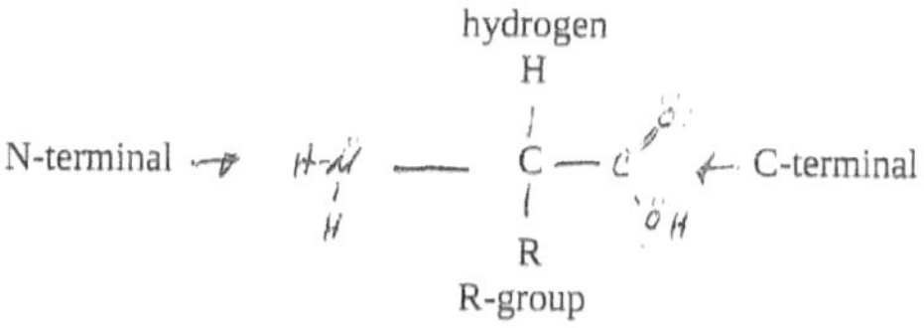

N-terminal → H–N –––– C – C ← C-terminal

hydrogen
H

R
R-group

The amine group is also called the N-terminal.

The central carbon has four different side chains and can bend polarized light.

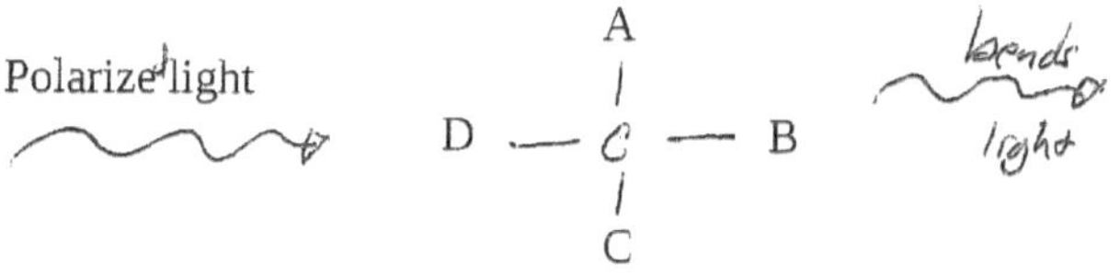

Polarize light

A
|
D — C — B
|
C

bends light

Every amino acid has a single hydrogen.
The carboxyl group is also known as the C-terminal.
Every amino acid has a side chain R which makes it unique.
Proteins are multiple amino acids linked together.

Carbohydrates are a group of compounds that provide energy. There are two basic groups: sugars and complex carbohydrates such as cellulose and starch.

The common carbohydrates are:

- Glucose (in honey)

  Fructose (in fruit)

  Galactose (in grains)

  Lactose (in milk)

  Sucrose (in sugar cane)

  Cellulose (60 percent of most plants)

  Starch (pastas, breads, rice, potatoes, and others)

Starch is similar to cellulose, but it forms an alpha helix instead of a straight line (linear).

Cellulose:　　　　　　　　　　　　　　　　　　(linear)

Starch:　　　　　　　　　　　　　　　　　　　(helix)

Cellulose fibers

Cellulose structure

Images

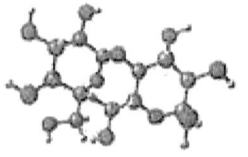 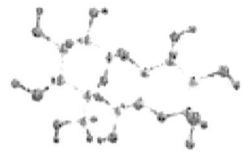 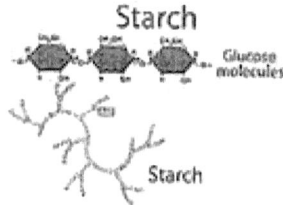

Hydrogen

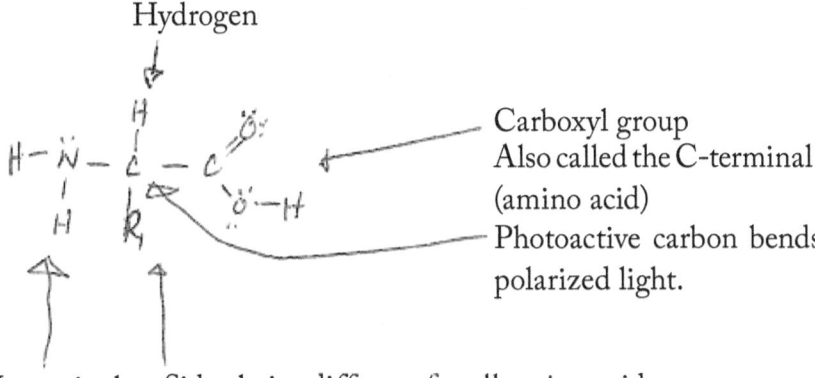

Carboxyl group
Also called the C-terminal
(amino acid)
Photoactive carbon bends
polarized light.

N-terminal: Side chain: different for all amino acids.
Also called
An amine
group.

N-terminal $H-N-C-C-N-C-C-N-C-C-N-C-C-O-H$ C-terminal

(protein)

Glucose (from honey)   Fructose (from fruit)

Galactose (from grains)

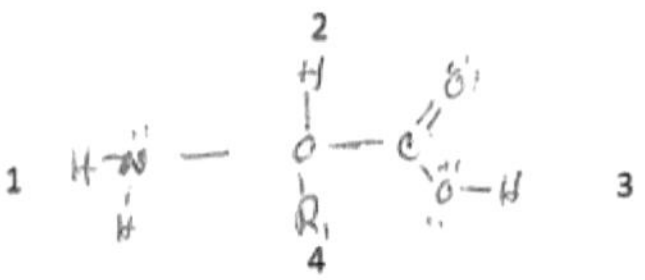

(amino acid)

1. N-terminal
2. Hydrogen
3. C-terminal
4. R group

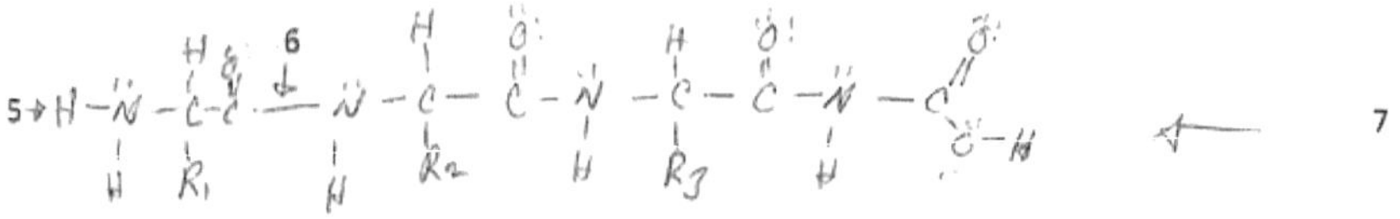

(protein)

5. N-terminal
6. Amide bond
7. C-terminal

Glucose (honey)   Fructose (fruits)

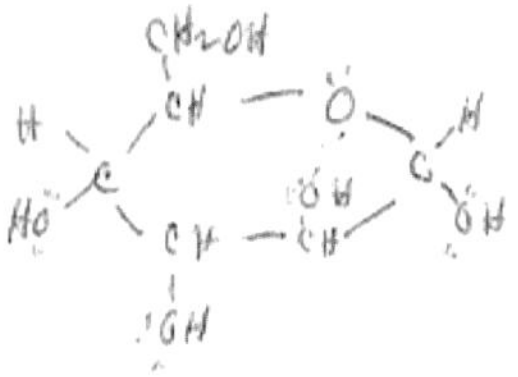

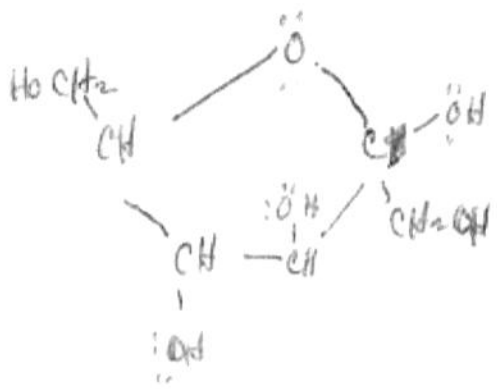

Galactose (grains)

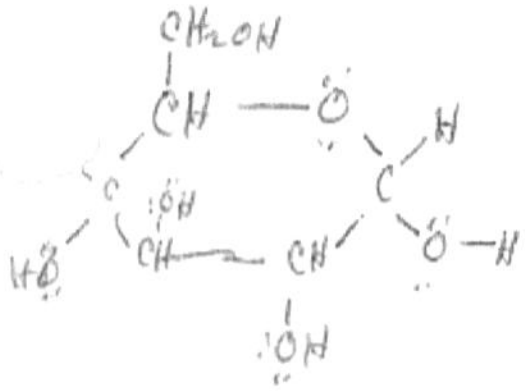

Note that galactose is similar to glucose, and that lactose is similar to maltose. The same atoms and the same number of atoms are just arranged differently.

Such chemicals are referred to as isomers. But even though they are very similar, enzymes that react with one will not react with the other.

Lactose (milk)

Maltose (grains)

Sucrose (sugarcane)

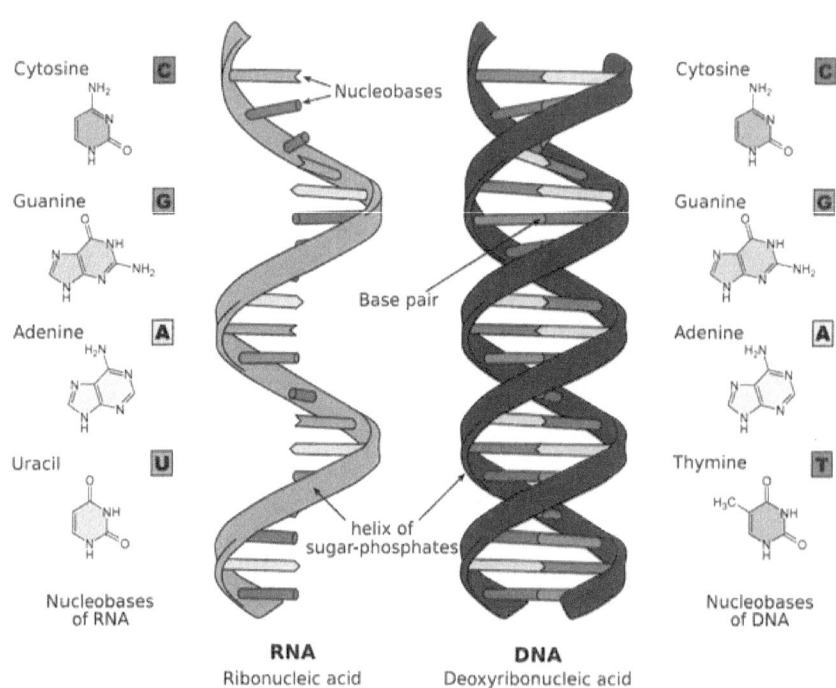

DNA

alpha helix (starch)

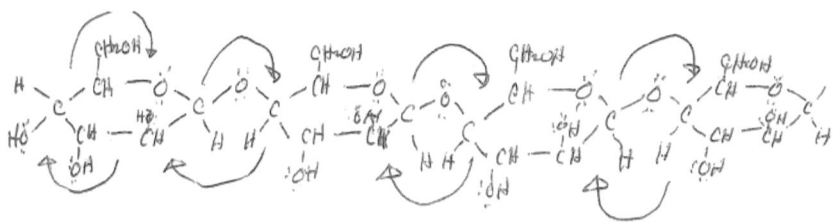

Starch (twists upon itself and forms a helix)

DNA

1 Nitrogen bases
2 Deoxyribose sugar
3 Phosphate links
4 Sense strand
5 Nucleosomes (made of eight histones)
6 Antisense strand

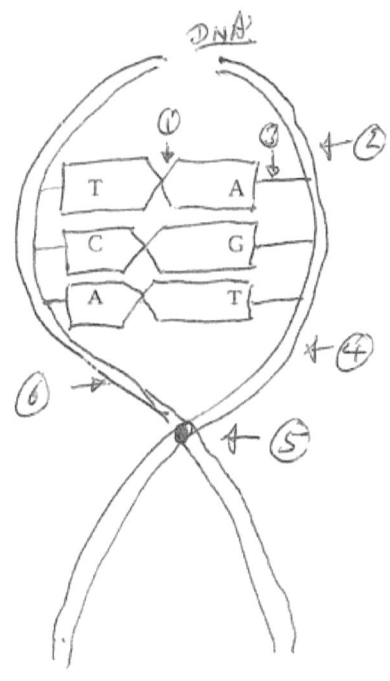

Messenger RNA: Unlike DNA, messenger RNA has only one strand. And instead of deoxyribose, it has ribose. The difference between them is that ribose contains one additional atom of oxygen.

Instead of thymine, messenger RNA uses uracil.

Transfer RNA:

- Has triplicate codons

  Transports amino acids

Ribosomal RNA is similar to messenger RNA but specifically codes for the function of the ribosomes.

Mitochondrial DNA directs the activity of the mitochondria.

When DNA opens, the sense and antisense strands are exposed. With the help of an enzyme called RNA polymerase, messenger RNA (mRNA) will form using the sense stand as a template:

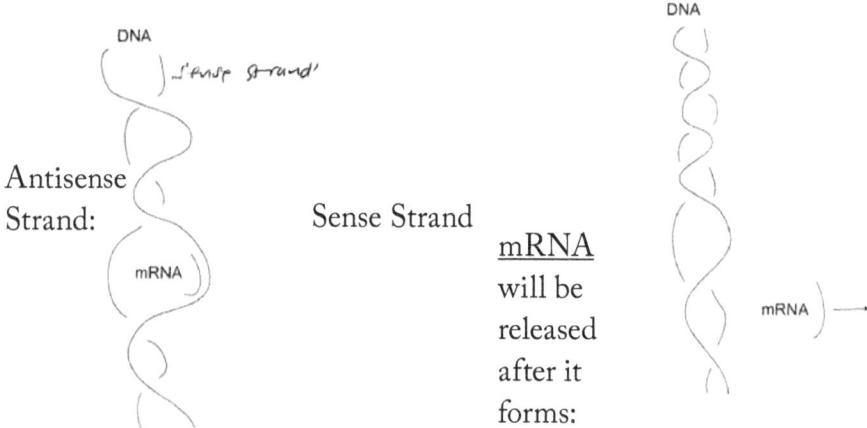

Antisense Strand: Sense Strand

mRNA will be released after it forms:

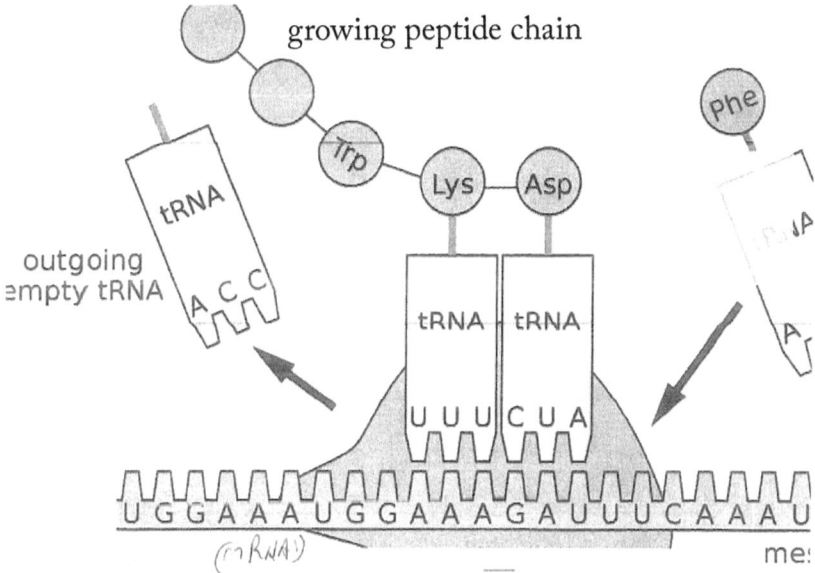

The interaction of tRNA and mRNA in protein synthesis.

Chemical Bonding

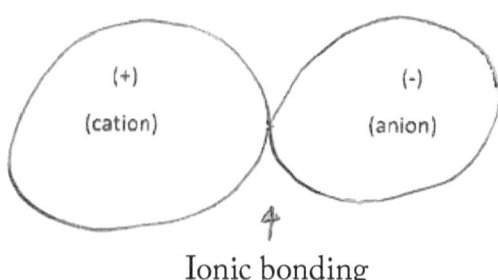

Ionic bonding

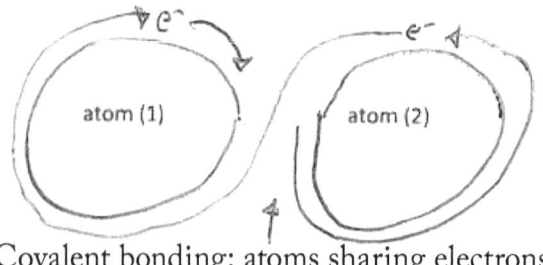

Covalent bonding: atoms sharing electrons

Single bond (covalent bonding)

covalent

Double bonding (covalent and a pi bond)

Triple bond (rare in biology)

$$H-C\equiv C-H$$

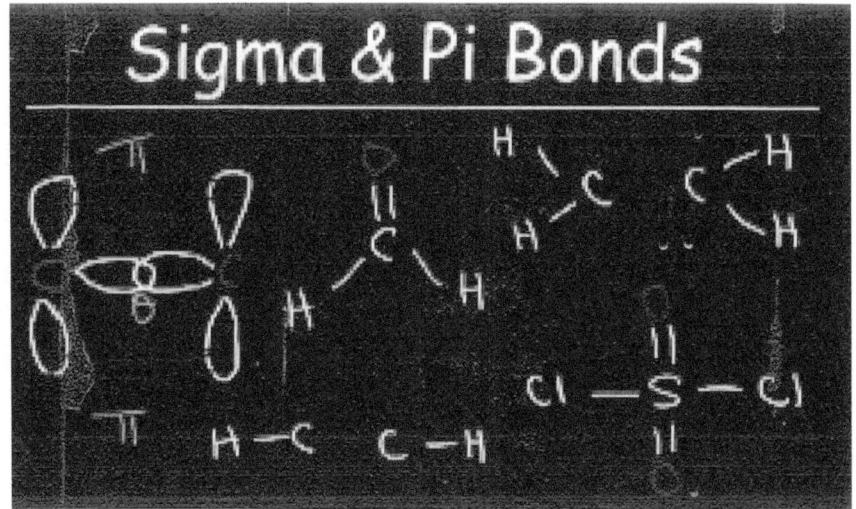

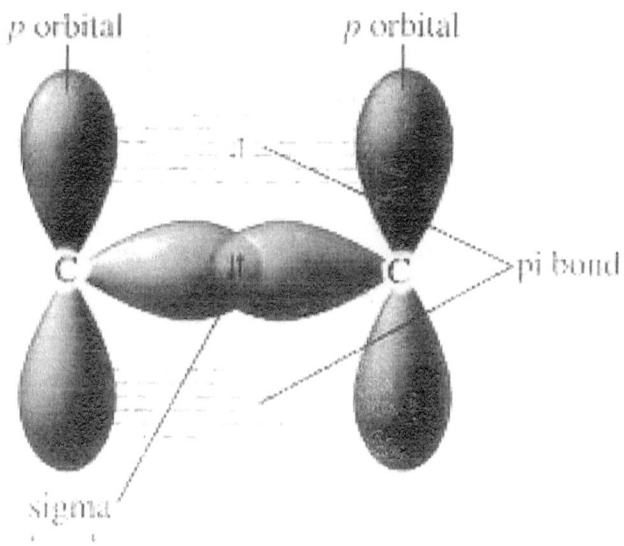

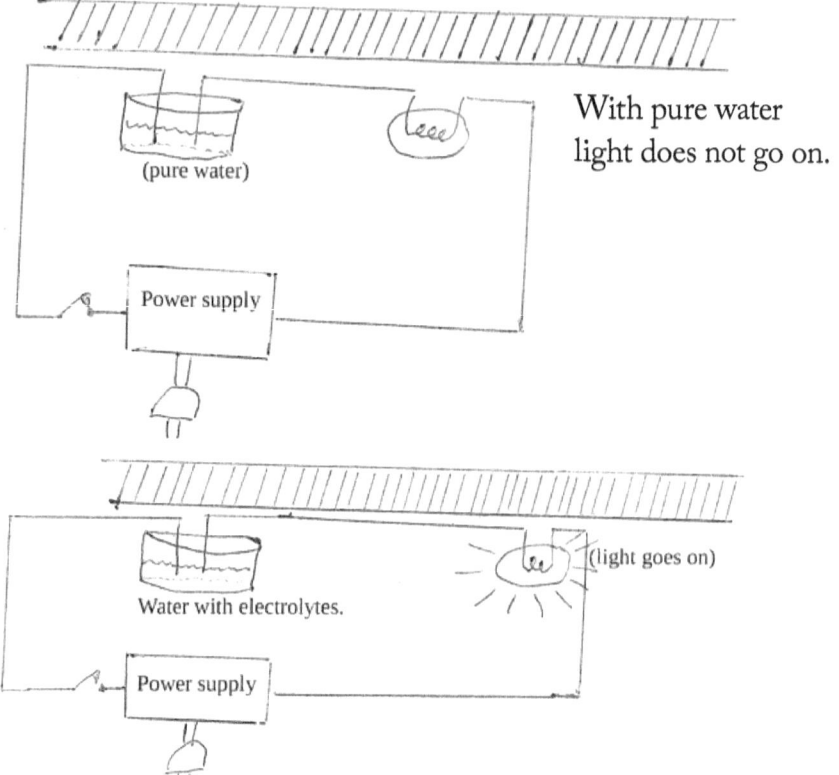

With pure water light does not go on.

(pure water)

Power supply

Water with electrolytes.

(light goes on)

Power supply

Electrolytes allow water to conduct electricity.

Niels Bohr

Planetary model of the atom:

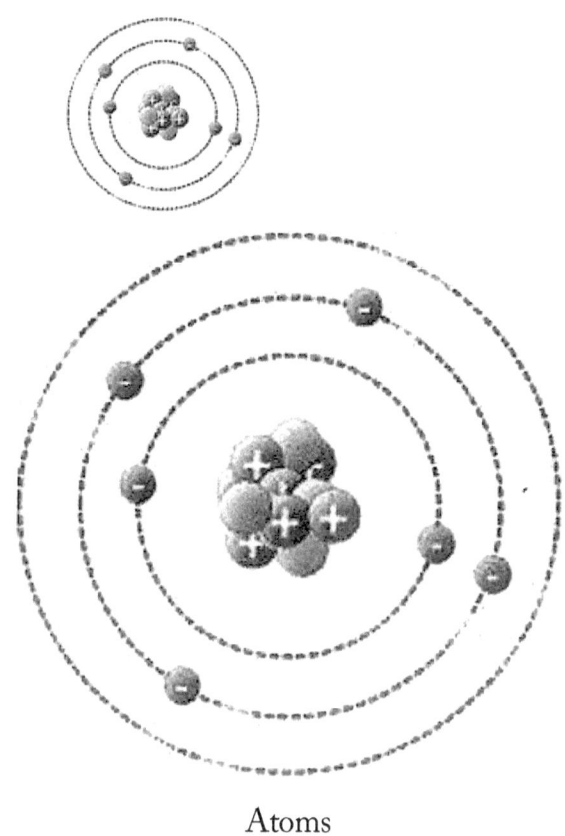

Atoms

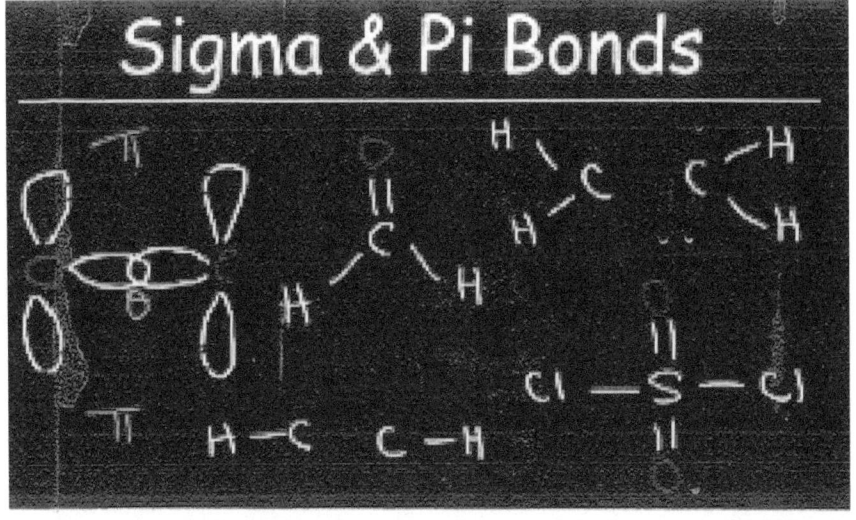

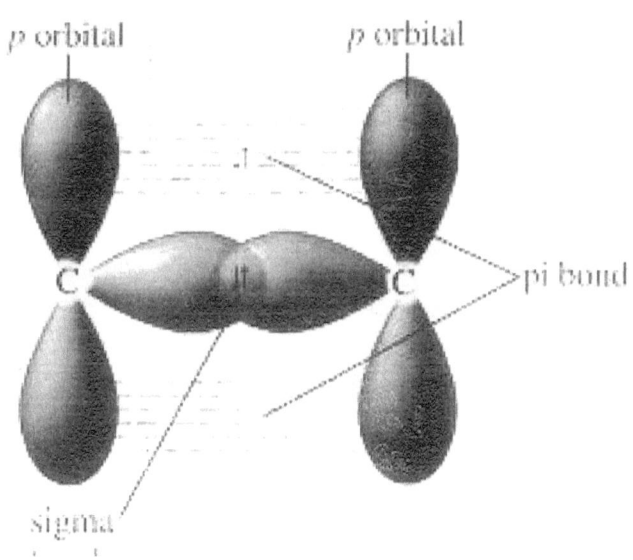

Radicals

Let us go on to discuss free radicals and ion radicals. Remember the Niels Bohr model of the atom? The nucleus is made of protons and neutrons, and electrons orbit around it.

Think of an atom as a miniature centrifuge with electrons orbiting the nucleus at almost the speed of light. Like all centrifuges, the system has to be balanced. Each electron has to be paired with another on the opposite side (separated by 180 degrees).

But supposing there were an odd number of electrons? This would be like an unbalanced centrifuge. It would be very unstable. To correct it the atom would either have to take on an electron or give one away. Remember, the centrifuge is running at almost the speed of light (186,252 miles/second).

Radicals are chemicals with an odd number of electrons in their valence shells (the outermost shell). They must either give an electron away or take one on. Most radicals take electrons away from other molecules. In the process they convert the molecule into a new radical, and the process repeats itself until a whole string of molecules has been damaged.

We need "radical sinks" that will give the radicals the electrons they need without becoming radicals themselves, such as vitamin A and vitamin E. This would stop radicals from inflicting damage on biological systems.

Suppose we create a radical by either adding or subtracting one electron from the valence shell? This would be an ion radical. If we add an electron, it would be an anion and a radical. If we subtract one, it would be a cation and a radical. These are called ion radicals because they have two reasons for either taking an electron or giving one away. They are radicals, and they are ions, and so they are even more reactive.

Radicals and ion radicals form within us all the time. We must maintain a supply of radical sinks to stop them from severely damaging biological systems. But there is one system that depends on them: lysozyme. Remember lysozyme? It is a very caustic substance that digests foreign invaders, damaged cells, and nutrients. It is a radical. This is what makes it so caustic.

Radical

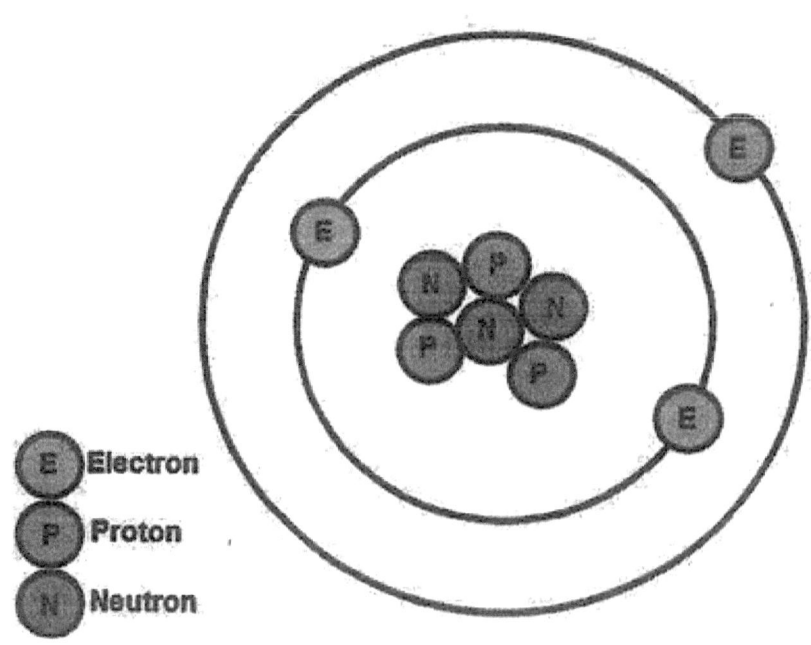

E Electron

P Proton

N Neutron

Radical Sinks:

radicals:

Vitamin A (radical sink)

Vitamin E (radical sink)

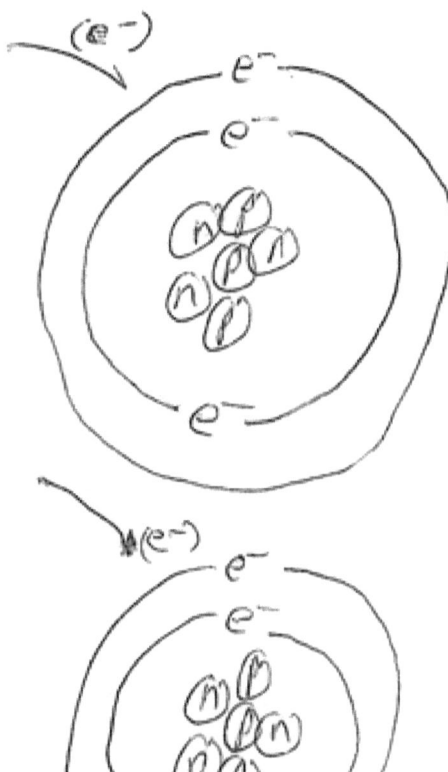

There are three ways for radical sinks to neutralize radicals.

1. The sink could be a very large molecule with a high atomic weight (amu). As the mass of the molecule increases, the effect of one unbalanced electron would diminish. It would be like taking a large centrifuge and removing one test tube. As the centrifuge gets bigger (and heavier) the effects of removing one test tube would diminish. It could be expressed mathematically as:

    Centripetal force of an unbalanced electron ($J_{(e)}$) is inversely proportional to the mass of the molecule (amu).

2. Another way to think about it: Electromotive force (emf) motivating an electron to migrate from one atom to another is directly proportional to the centripetal force ($J_{(e)}$) acting upon it.

$$emf \; \alpha \; J.$$

Centripetal force (J) is equal to 1/2 m x w squared/radius

In this case angular velocity (W) is almost the speed of light (186,252 miles/second).

Centripetal force (J) is equal to 1/2 m x w squared/radius

Therefore:

The mass of the electron and the speed of light are constants. They are not going to change. So as radius (r) gets bigger, electromotive force (the force that motivates a radical to either take or give away) diminishes.

as radius gets bigger emf gets smaller

3. Remember that a valence shell with an unbalanced electron could achieve stability by either taking an electron on or by giving one away. Suppose we put two radicals together, and one takes an electron and the other gives it away. They would neutralize each other.

So now we have one final question. If there is an unbalanced electron, how does the atom decide what it will do? Will it give one away or take one on? The answer is the "atomic number" (the number of protons in the nucleus). Remember, protons have a positive charge and attract electrons. If an atom has a large number of protons (high atomic number), it will stabilize the radical by taking an electron on (protons attract electrons because electrons have a negative charge and protons have a positive). If an atom attracts electrons, we say that it is "electronegative." If an atom has a small atomic number (few protons in its nucleus), it would not attract electrons and it would stabilize its valence shell by giving electrons away (such an atom would be "electropositive").

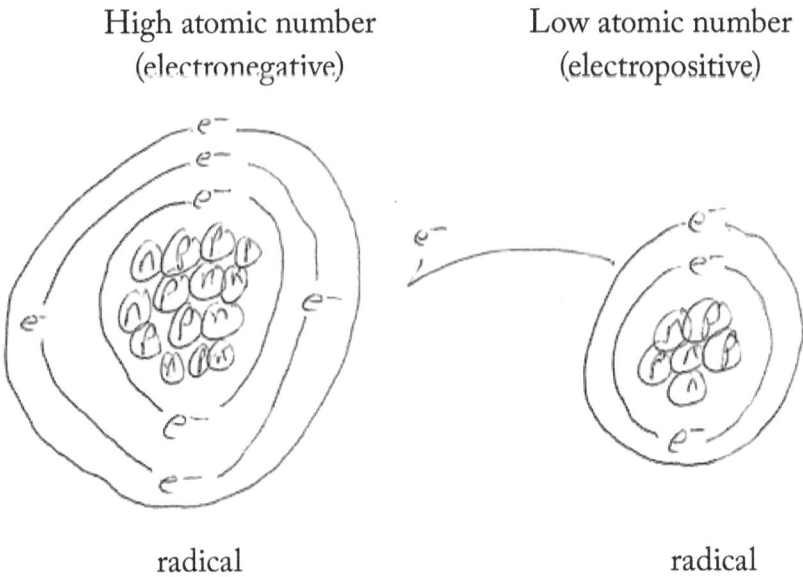

High atomic number (electronegative)   Low atomic number (electropositive)

radical   radical

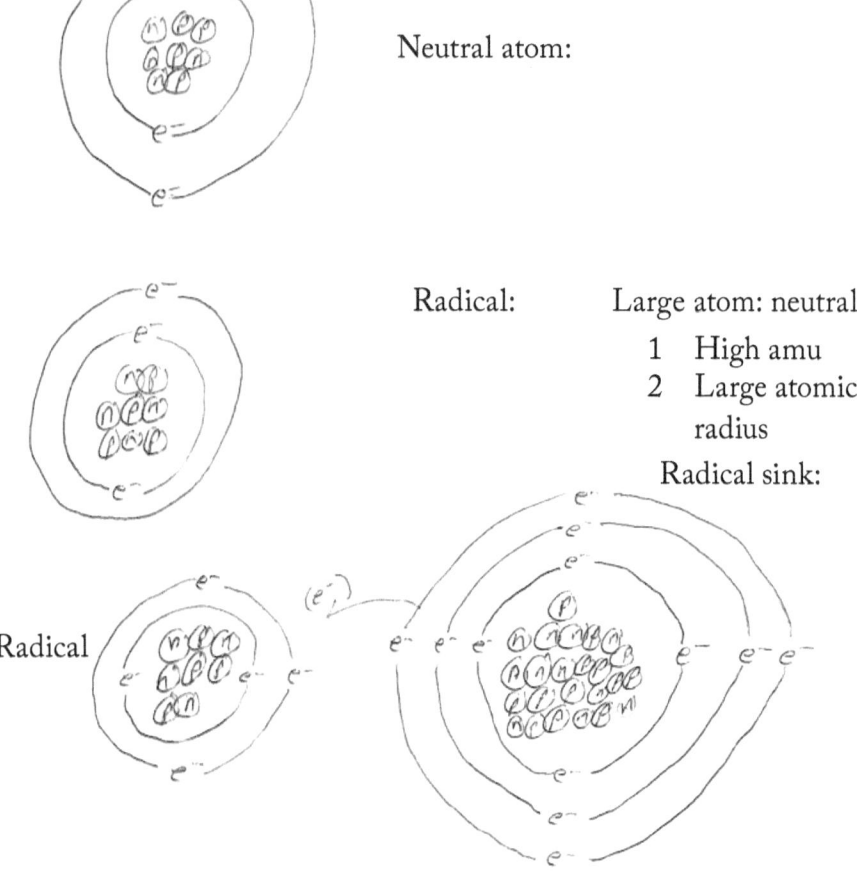

Neutral atom:

Radical:

Large atom: neutral
1   High amu
2   Large atomic
    radius

Radical sink:

Radical

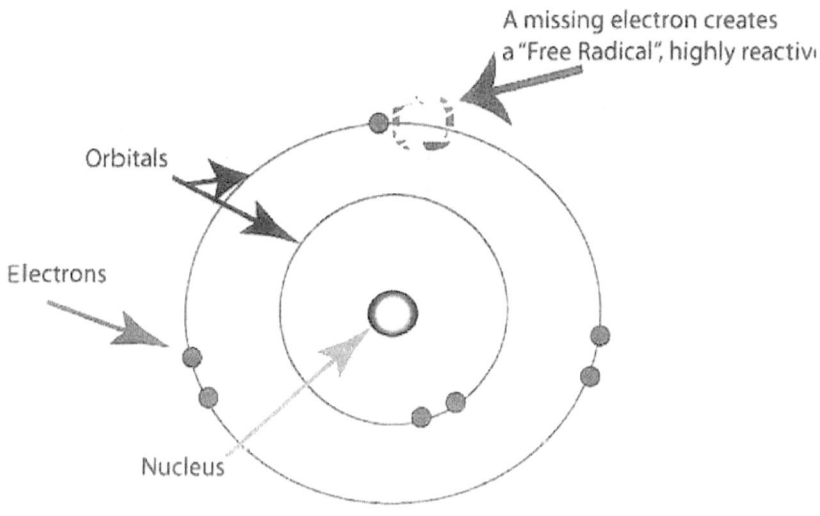

A missing electron creates
a "Free Radical", highly reactive

Orbitals

Electrons

Nucleus

Free Radical

Radical

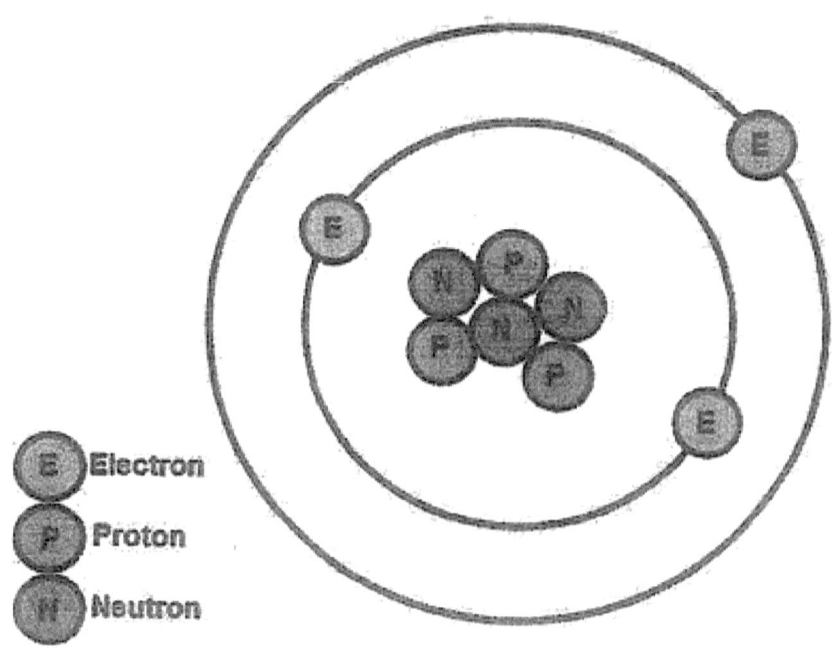

Radical Sinks:

radicals:

Vitamin A (radical sink)

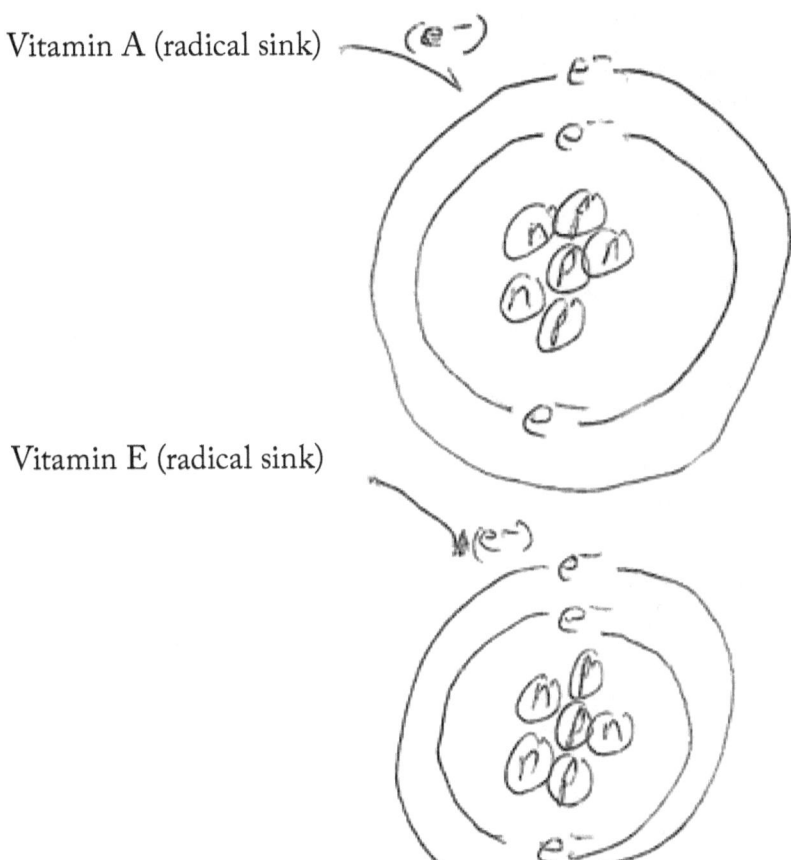

Vitamin E (radical sink)

Moses Gomberg
1900

Discovered Free Radicals

Let us discuss three of the most important processes in biology:

1. Osmosis
2. Equilibrium
3. Diffusion

Osmosis and Diffusion

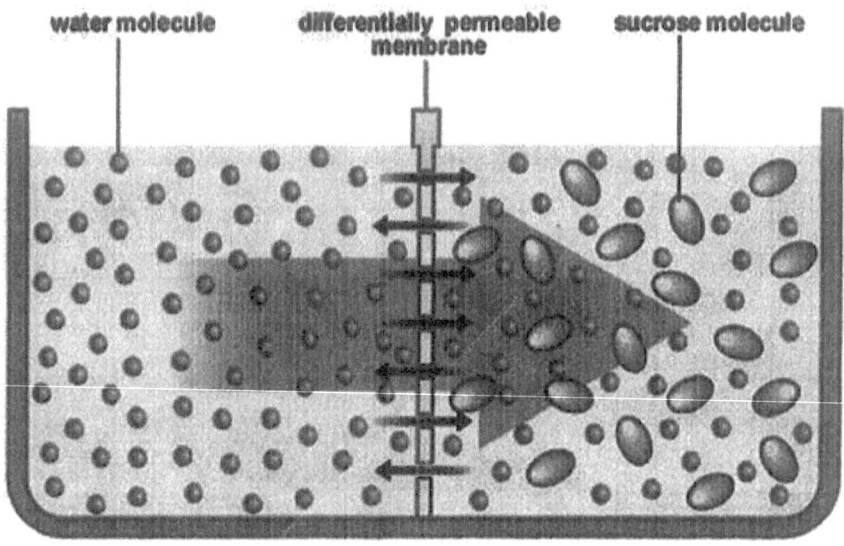

water molecule   differentially permeable membrane   sucrose molecule

When substances dissolve in a fluid (liquid or gas), they will equally distribute themselves throughout available space. This is because the molecules of fluid are in motion, a phenomenon known as "kinetic heat" and also referred to as "Brownian motion." Molecules of fluid and solute will be in motion and will equally distribute themselves throughout available space.

Let us say that we have a tank filled with water, and it is divided by a "semipermeable" membrane. Then we add sugar to half the tank.

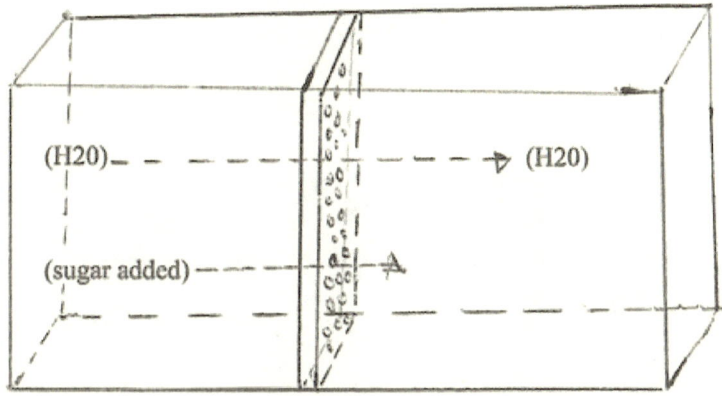

"Semipermeable" means that the divider is porous (contains holes). The pores will allow molecules of water and sugar to distribute themselves throughout the tank.

1. The movement of water is osmosis.
2. The movement of sugar is diffusion.
3. The ability of water and sugar to equally distribute themselves is equilibrium.

Osmosis and diffusion make equilibrium possible, and this is how cells obtain oxygen, water, and nutrients. It also allows them to expel carbon dioxide and waste products.

Acid-Base Theory

Dr. Svante Arrhenius won a Nobel Prize for describing the activity of Arrhenius acids.

Hydrochloric acid + water → hydrogen ion + chloride + water

1884

Let us talk about strong and weak acids and bases:

Strong acids include:

Nitric acid

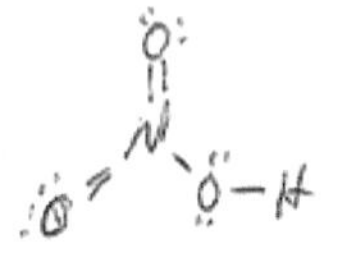

Hydrochloric acid

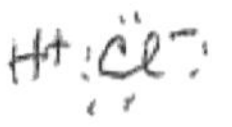

Sulfuric acid

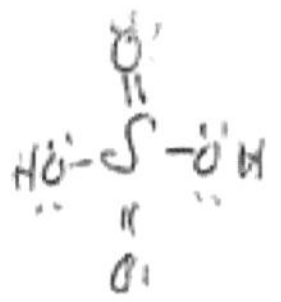

If an acid does not dissociate 100 percent at room temperature in water, it is a weak acid. This is the only criterion. But some strong acids are stronger than others. The more easily an acid dissociates, the stronger it is, and that depends on the stability of the anion. As the anion becomes more stable, the acid will dissociate with greater ease and will be stronger.

HA	Water	H^+		A^-
Acid		hydrogen ion	$+$	anion

Hydrochloric acid, nitric acid, and sulfuric acid are stronger acids. But sulfuric is stronger than hydrochloric, and hydrochloric is stronger than nitric because sulfate is more stable than chloride and chloride is more stable than nitrate.

Examples of weak acids (that do not dissociate 100 percent at room temperature) include:

Weak acids:

Acetic acid $CH_3 - C$ ...

Citric acid

Carbonic acid

Lactic acid

Tartaric acid

Hydrofluoric acid

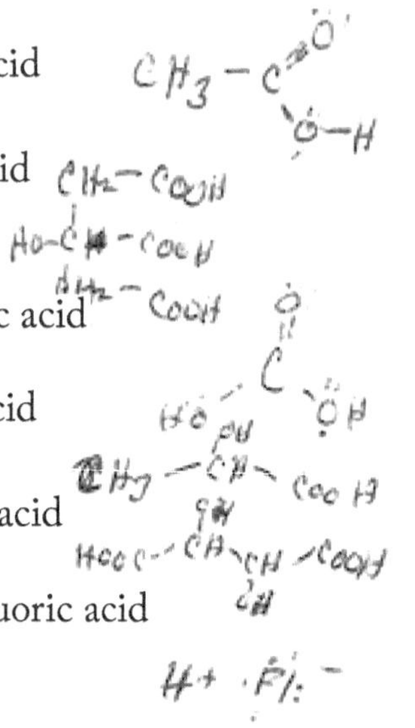

Examples of strong acids include:

Nitric acid

Hydrochloric acid

Sulfuric acid

Chloric acid

Perchloric acid

Chromic acid

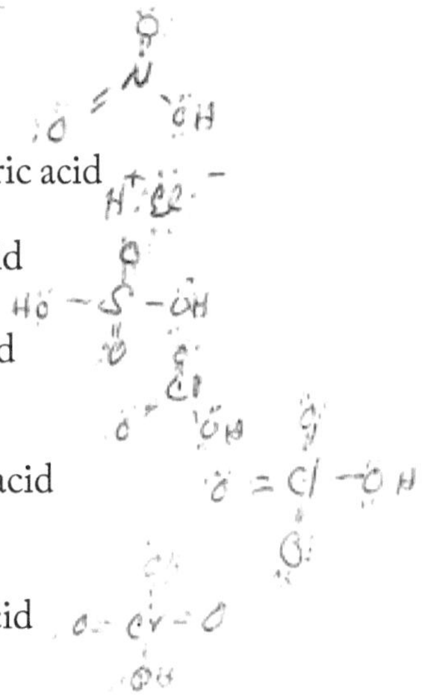

So let us discuss acid-base theory. Acid-base theory states that strong acids and bases interact to form weaker acids and bases.

How do we measure the strength of an acid? With pH. pH = -log [H^+]. Neutral water has a pH of 7. Bases have a pH that is greater than 7. The stronger the base, the higher the pH, and acids have a pH of less than 7. The lower the pH, the stronger the acid. When discussing bases, some chemists speak of pOH:

pOH = -log [OH^-]

So what is a Lewis acid? A Lewis acid is a substance that does not dissociate into an anion and a hydrogen ion when mixed with water: but it will cause the pH to drop below 7. The way it works is this: at room temperature one molecule of water in every 10 million will dissociate into a hydrogen ion and a hydroxide ion:

$$H-\overset{..}{\underset{..}{O}}\diagdown_H \rightarrow H^+ + OH^-$$

If you took a flask of triple distilled water, one molecule in 10^7 would do the above. This would be represented mathematically as pH=-log ($1/10^7$). The pH of pure water is 7.

Suppose we add a salt such as copper II sulfate. The copper II would combine with the hydroxide ion and cause the pH to drop below 7. So the pH of pure water is 7. The copper 2^+ ion would combine with the OH^- to form copper II hydroxide [Cu (OH)$_2$]. But the sulfate would not combine with the H^+, and the excess H^+ would cause a drop in pH.

$$H-\overset{.}{O}\diagdown_H \rightarrow \quad \left(\begin{array}{c} OH^- \\ Cu^{2+} \end{array}\right) \quad + H^+$$
(water)

$$CuSO_4 \rightarrow Cu^{2+} \qquad + SO_4^{2-}$$
(copper II sulfate)

$$Cu^{2+} + 2 OH^- \rightarrow Cu(OH)_2\downarrow \; (precipitates)$$

Examples of Lewis acids include:

Copper II sulfate $\ddot{O} = \overset{\overset{\displaystyle :\ddot{O}:^-}{|}}{S} = \ddot{O}:$ Cu^{2+}

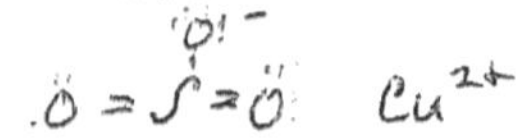

Calcium chloride $Ca(Cl)_2$

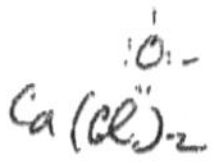

Silver nitrate $AgNO_3$

Iron nitrate $FeNO_3$

If a substance were to bind hydrogen ions, it would raise pH and would bring the pH of water above 7. That would be a Lewis base, and these include:

Sodium acetate $CH_3 - C\overset{\displaystyle =\ddot{O}:}{\underset{\displaystyle \ddot{O}: - Na^+}{}}$

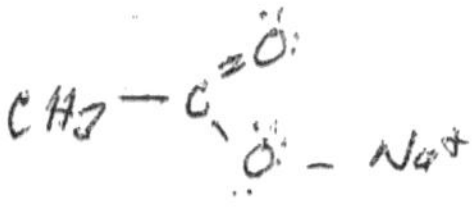

Lithium carbonate

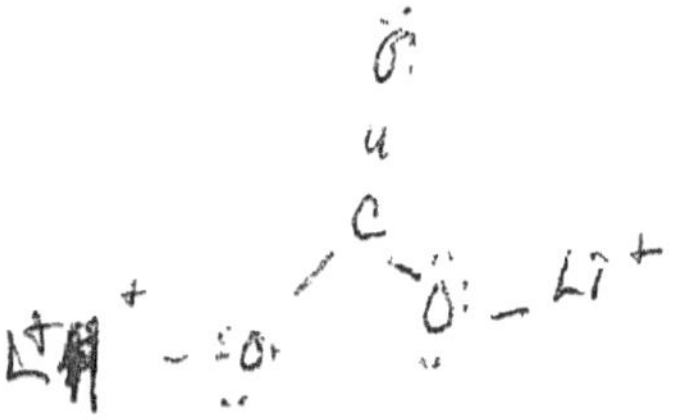

$Li^+ \quad -:\ddot{O}: \qquad \overset{\displaystyle \ddot{O}:}{\underset{\displaystyle C}{}} \quad \ddot{O}: - Li^+$

106

Potassium citrate

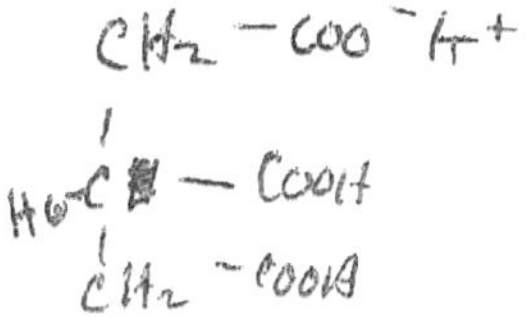

$CH_2 - COO^- K^+$
$|$
$HO-C - COOH$
$|$
$CH_2 - COOH$

Potassium cyanide $K^+ \quad :C \equiv N:$

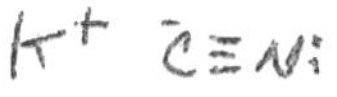

Arrhenius and Lewis acids are common in biology.

103

24)

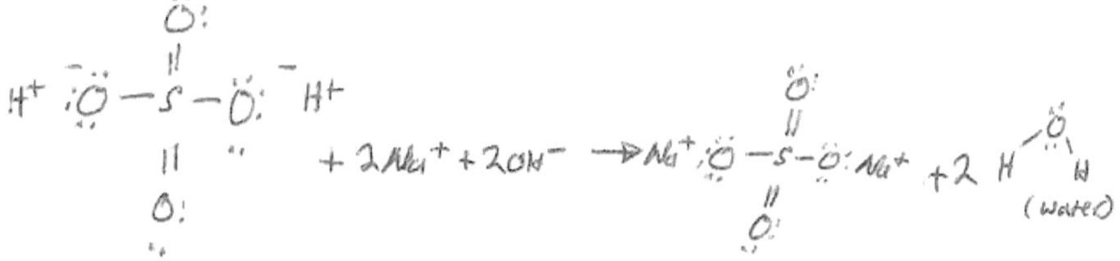

Sulfuric acid Sodium hydroxide Sodium sulfate

$$pH = -\log[H^+]$$

$$pOH = -\log[OH^-]$$

Now add

$$H{-}O{-}H \rightarrow OH^- + H^+ \qquad Ca^{2+} \quad SO_4^{2-}$$

$$2OH^- + Ca^{2+} \rightarrow Ca(OH)_2\downarrow$$

but: $2H^+ + SO_4^{2-} \not\rightarrow 2H^+ + SO_4^{2+}$

Sodium acetate:

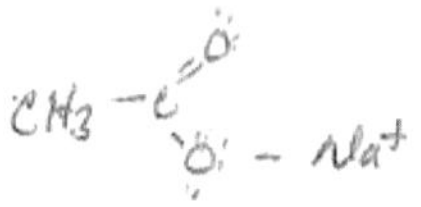

Lithium carbonate:

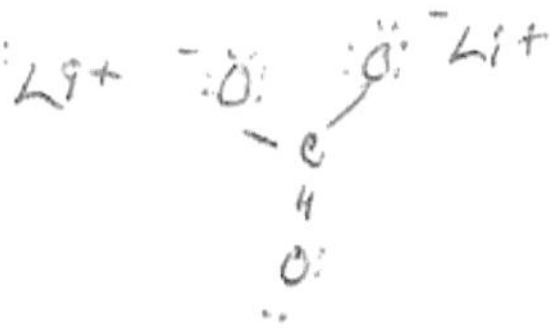

Potassium cyanide:

$$K^+ \quad {}^-C\equiv N:$$

1. Vinegar (acetic acid)

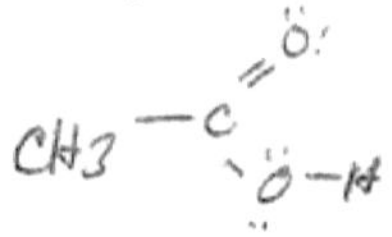

$$CH_3 - C \overset{= \ddot{O}:}{\underset{\ddot{O} - H}{}}$$

2. Carbonic acid (makes soda pop bubbles)

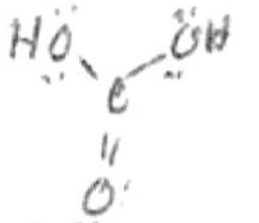

$$H\ddot{O} \underset{\underset{\ddot{O}:}{\overset{\|}{}}}{\diagdown} C \diagup \ddot{O}H$$

3. Citric acid (makes lemons sour)

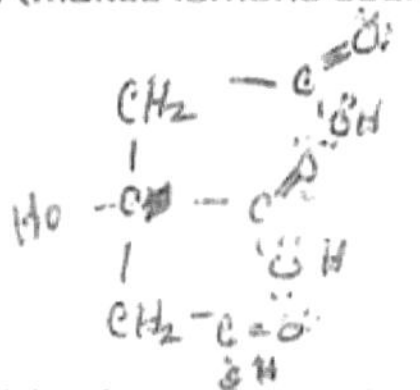

$$CH_2 - C \overset{=\ddot{O}:}{\underset{\ddot{O}H}{}}$$
$$HO - C - C$$
$$\ddot{O} H$$
$$CH_2 - C = \ddot{O}$$
$$\ddot{O} H$$

4. Lactic acid (makes milk sour)

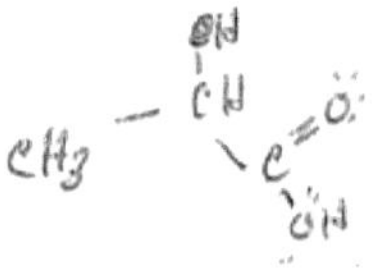

(2 hydroxy propionic acid)

$$CH_3 - \overset{\overset{\ddot{O}H}{|}}{CH} \diagdown C \overset{=\ddot{O}:}{\underset{\ddot{O}H}{}}$$

5. Hydrofluoric acid

$$H^+ + :\ddot{F}:^-$$

The final category are the Bronsted-Lowery acids. They include all oxidizing agents that take electrons away from other molecules. They are rarely seen in biology. Examples include:

Hydrogen peroxide

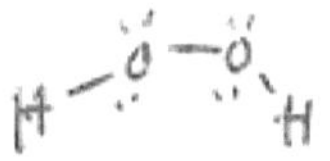

Potassium permanganate

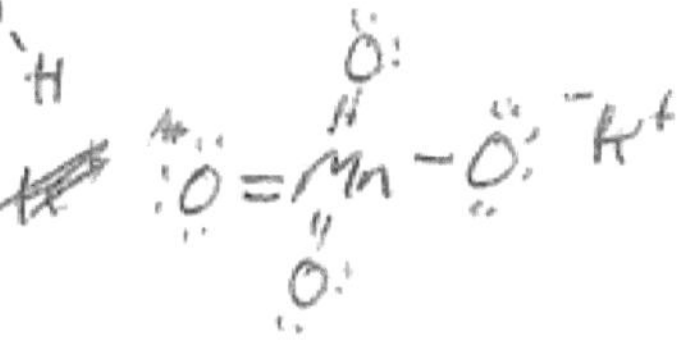

Ozone:

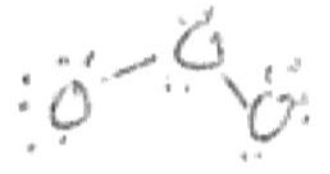

Sodium nitrate

A Bronsted-Lowery base would give an electron away. Examples include:

Hydrogen gas H_2
Aluminum   Al
Magnesium   Mg
Carbon   C
Sulfur   S

Let Us Briefly Revisit the Topic of Lewis Acids

Additional examples include:

Ketones
Silver Nitrate
Iron Nitrate

Enzyme Activity as a Function of pH

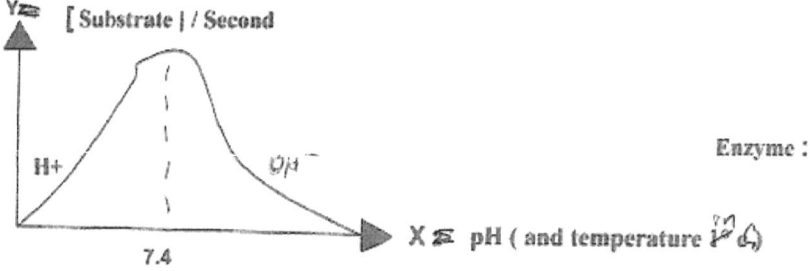

And let us discuss the solubility laws. There are seven of them:

- All salts made from elements in column 1A of the periodic table will dissolve in water at room temperature—72 degrees F, or 22.2 degrees C.

  All bases made from elements in column 1A will dissolve in water at room temperature.

  All bases made from elements in column 2A will dissolve in boiling water only (212 degrees F or 100 degrees C)

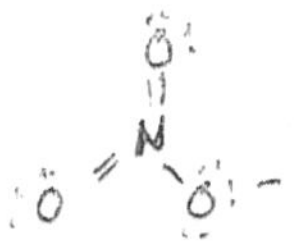

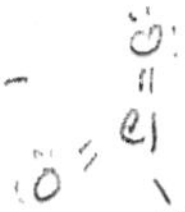

All nitrates are water-soluble at room temperature.

• All chlorates are water-soluble at room temperature.

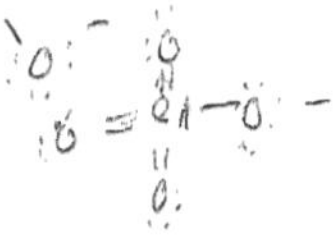

All perchlorates are water-soluble at room temperature

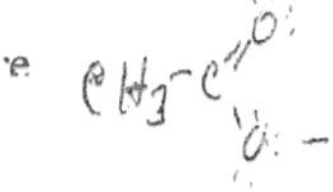

All acetates are water-soluble at room temperature.

Column 1A includes:

Potassium

Sodium

Lithium

Francium

Rubidium

Cesium

Column 2A includes:

• Magnesium

Calcium

Strontium

Beryllium

Barium

Radium

Substances that dissolve in water at room temperature are 100 percent soluble at body temperature (98.6°F, 37°C),

There is one more topic that we should discuss before leaving chemistry: the composition, structure and activity of enzymes.

Enzymes are biological catalysts. They promote chemical reactions in biological systems without being consumed by them.

Enzymes are proteins, which contain cofactors, such as trace metals, and coenzymes (usually synthesized from vitamins).

When systems create enzymes, they link amino acids together with amide bonds. In doing so, they form a long protein polymer (a long molecule which is a protein). This molecule will twist upon itself to form an alpha helix, a springlike structure as seen on X-ray crystallography. The alpha helix will then twist upon itself again:

$$\begin{array}{ccccccc}
\text{H} & & \text{O} & & \text{H} & & \\
| & & || & & | & & \text{O} \\
-\text{C}- & & \text{C}- & \text{N}- & \text{C}- & \text{C} & \\
| & & & | & | & & \text{O-H} \\
\text{R}_1 & & & \text{H} & \text{R}_2 & &
\end{array}$$

In the end we get a long protein molecule that is folded upon itself many times. There will be a distribution of partial electric charges upon this molecule that are referred to as moments:

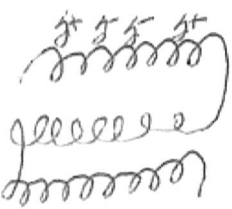

A long protein molecule that forms an alpha helix like a spring. Then it folds upon itself.

Pka, Pkb, and Solubility:

Let us take a few minutes to discuss pH and Pka. Pka is the solubility coefficient for substances mixed with acids. Pka is equal to negative log base 10 of [cation] [anion] / [undissociated salt] in an acid solution. pH < 7.

$$Pka = -Log \frac{[cation] \cdot [anion]}{[C \cdot A]}$$

In an acid

As the solubility of the salt diminishes (as it becomes less water-soluble), the Pka increases. So there is an inverse relationship. As solubility goes down, Pka goes up, and vice versa. Examples of substances with high Pka values include:

$$Pka \propto \frac{1}{Solubility}$$

Oyster Shell Calcium	$CaCO_3$
Iron Sulfate	$Fe_2(SO_4)$
Aluminum Hydroxide	$Al(OH)_3$

Pkb is the same concept but in water that has a base pH (above 7). Examples of substances with high Pkb values include:

$$Pkb = -Log \frac{[cation] \cdot [anion]}{[CA]}$$

In a base

Magnesium carbonate	$MgCO_3$
Calcium carbonate	$CaCO_3$
Silver carbonate	Ag_2CO_3

Remember, high Pka and Pkb means that very little of the substance dissolves.

$$Pka = -Log\,(10) \quad \frac{[anion] \times [cation]}{[undissociated\ salt]} \qquad \text{In an acid solution.}$$

$$Pka = -Log\,(10) \quad \frac{[A] \times [C]}{[A.C]} \qquad \text{In pH <7 (acid)}$$

$$Pkb = -Log \frac{[A] \times [C]}{[A.C]} \qquad \text{In pH > 7 (base)}$$

$$Pka \; \alpha \; \frac{1}{Dissolved\ salt}$$

$$\frac{Pkb}{} \; \alpha \; \frac{1}{Dissolved\ salt}$$

Buffers

In biology we have substances known as buffers. By definition they resist change in pH even if you dilute them, dehydrate them, or mix them with an acid or a base. Blood is a buffer, and the pH of the human body stays at 7.4, which is critical to the function of enzymes. If we plot moles of substrate metabolized per second as a function of pH, we get the following curve:

Moles/second metabolized:

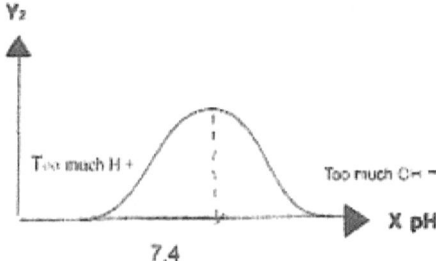

As the pH goes below or above 7.4, the desired activity drops. The reason for this is as follows: remember that enzymes have a positioning site and a catalytic site. Both depend upon a distribution of partial electric changes known as moments.

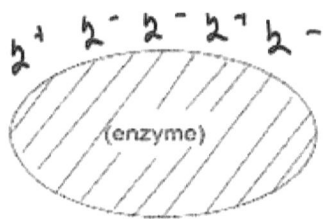

If there is an excess of either hydrogen or hydroxide ions (low pH or high pH), the distribution of these charges would be disturbed.

[moles] / second.

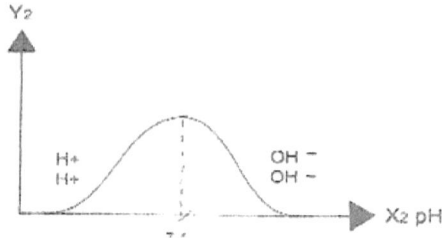

It is the job of the buffer (along with the lungs and the kidneys) to hold pH constant. Here's how buffers work: suppose we take phosphoric acid:

$$\ddot{O}: \quad\quad\quad\quad\quad\quad\quad\quad \ddot{O}!$$
$$H\ddot{O} - \underset{\underset{:OH}{|}}{\overset{||}{P}} - \ddot{O}H \longrightarrow H\ddot{O} - \underset{\underset{:OH}{|}}{\overset{||}{P}} - \overset{(-)}{\ddot{O}!}$$

phosphoric acid $\quad H^{(+)}$ $\quad$ di-hydroxy ion

Let us expose the di-hydroxy ion to four possible conditions:

$$\ddot{O}: \quad\quad\quad\quad\quad\quad\quad\quad \ddot{O}:$$
$$H\ddot{O} - \underset{\underset{\cdot OH}{|}}{\overset{||}{P}} - \ddot{O}^{-} \quad + \quad H^{+} \rightarrow H\ddot{O} - \underset{\underset{'OH}{|}}{\overset{||}{P}} - \ddot{O}H$$

di-hydroxy phosphate $\quad$ (acid) $\quad$ phosphoric acid

Add an acid. The di-hydroxy ion will combine with it and neutralize it.

$$:O! \quad\quad\quad\quad\quad\quad\quad :O:$$
$$H\ddot{O} - \underset{\underset{:OH}{|}}{\overset{||}{P}} - \ddot{O}^{(-)} \quad + \quad :OH \rightarrow H\ddot{O} - \underset{\underset{:O^{(-)}}{|}}{\overset{||}{P}} - \ddot{O}^{(-)}$$

(base) $\quad$ monohydroxy phosphate $\quad$ + H2O

Add a base. The di-hydroxy ion will release a hydrogen ion and neutralize it.

$$:O! \quad\quad\quad\quad\quad\quad\quad :O!$$
$$H\ddot{O} - \underset{\underset{'OH}{|}}{\overset{||}{P}} - \ddot{O}^{(-)} + H2O \longrightarrow H\ddot{O} - \underset{\underset{O^{-}}{|}}{\overset{||}{P}} - \ddot{O}^{(-)} \quad + \quad H^{+} + H2O$$

monohydroxy phosphate

Add water (H2O): the di-hydroxy ion will release a hydrogen ion and maintain pH.

$$:'O \quad\quad\quad\quad\quad\quad\quad \ddot{O}'$$
$$H\ddot{O} - \underset{\underset{:OH}{|}}{\overset{||}{P}} - \ddot{O}^{(-)} \; -H2O \rightleftharpoons H\ddot{O} - \underset{\underset{'OH}{(}}{\overset{||}{P}} - \ddot{O}H$$

$\quad\quad\quad\quad\quad\quad H^{+}$

di-hydroxy phosphate $\quad\quad\quad\quad\quad$ phosphoric acid.

Take water away: the di-hydroxy ion will combine with a hydrogen ion and maintain pH.

And our last category (which is very common in biology) are the acid anhydrides: chemicals that form Arrhenius acids when dissolved in water. Examples:

Carbon dioxide

Sulfur dioxide

Hydrogen chloride

Sulfur trioxide

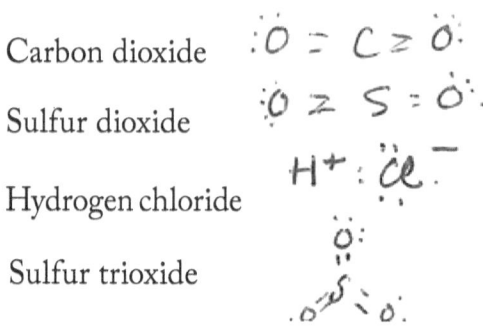

As pH changes (either up or down), enzyme metabolism changes dramatically. If pH does not stay at a constant 7.4, the rate of catabolism will drop sharply.

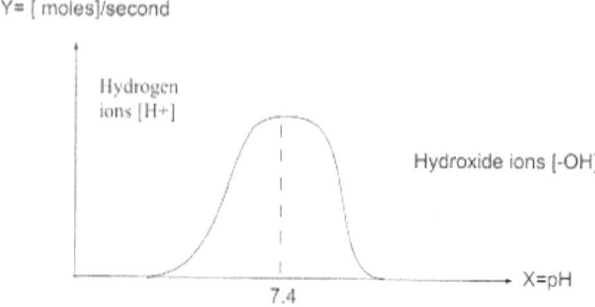

The same is true of temperature. If the temperature of the human body goes above or below 98.6°F, 37°C, metabolism will drop off.

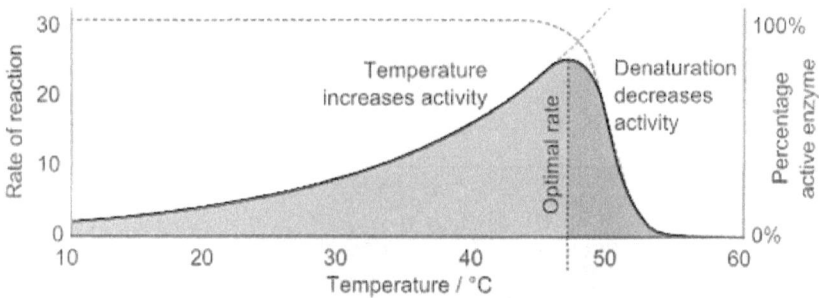

Blood is not the only buffer. Many cells act as buffers.

In addition to enzymes, pH plays another role. It affects the solubility of other substances. Acids, acid anhydrides, and Lewis acids will dissolve more readily in water that has a pH greater than 7, whereas sodium bicarbonate is less soluble.

Acetic acid: $CH_3-C{\overset{O}{\underset{OH}{}}}$

Carbon dioxide: $O=C=O$

Calcium chloride: $CaCl_2$

} More soluble in water with a base pH greater than 7.

Sodium bicarbonate:

Less soluble in water with a pH greater than 7.

Let us shift our attention to genetics and inheriting characteristics from our parents.

At this point we will discuss the two forms of cell division, mitosis and meiosis.

Mitosis is the process by which a cell produces an exact replica. This is key to growth, healing, and replacing worn-out cells. Typical examples include:

- Healing of an injury

    Replacing dead skin

    The growth of bones during childhood

During mitosis each chromosome splits to form two identical replicas known as chromatids.

The chromosomes divide first:

This is mitosis (m). Once complete the cell divides. The two new cells are referred to as daughter cells. The daughter cells then grow in what is known as the growth one or the G1 phase.

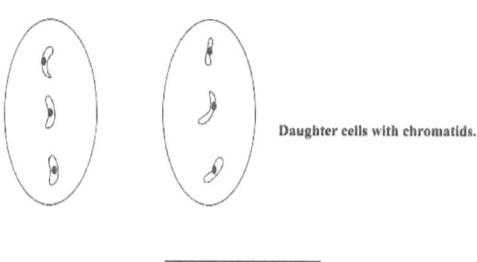

Daughter cells with chromatids.

Mitosis

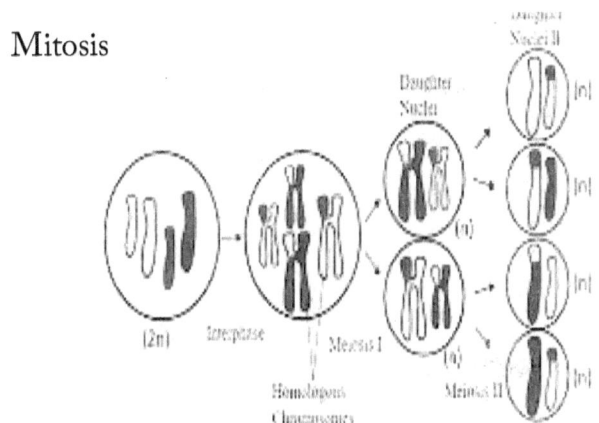

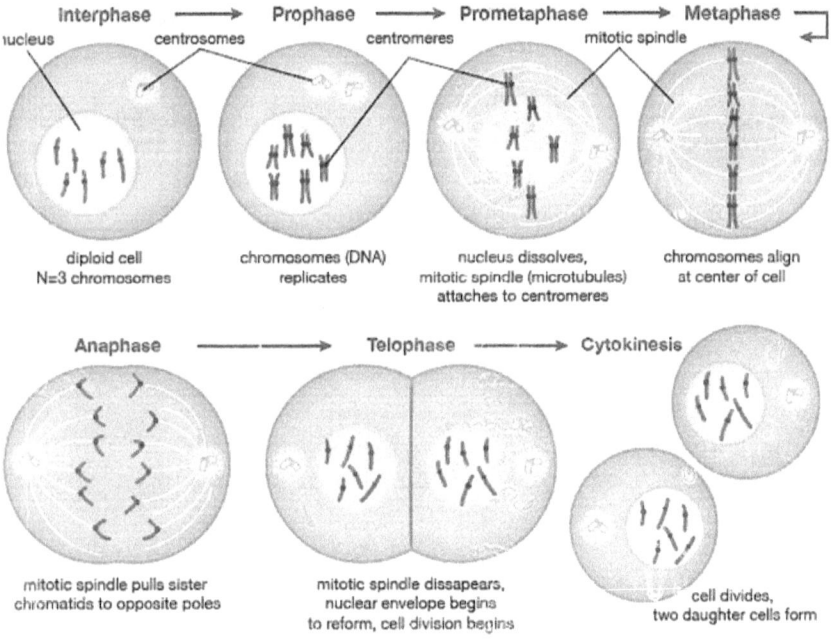

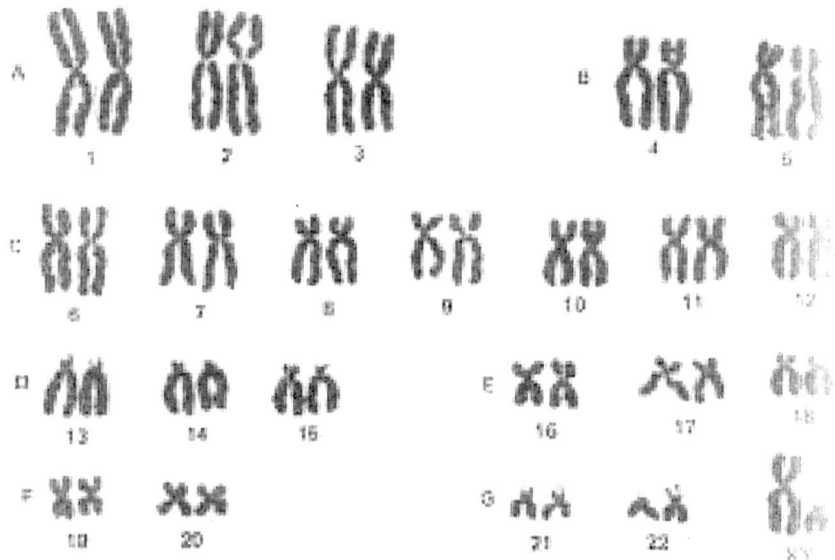

Male karyotype

Female Human Karyotype

When chromosomes are laid out in a row, the pattern is referred to as "karyotype."

Chromosomes have two arms. The Q arm is the longer of the two, and the P arm stands for "petit."

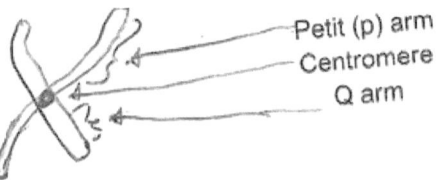

Chromosomes come in pairs. We have 23 pairs in our cells.

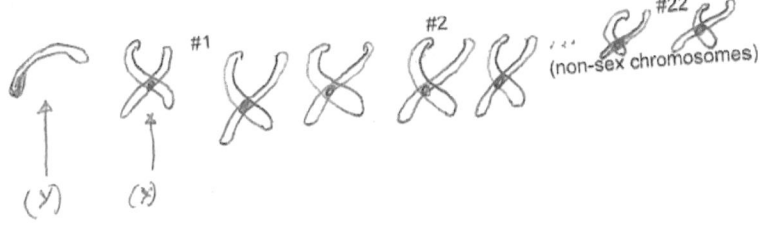

Male Human Karyotype

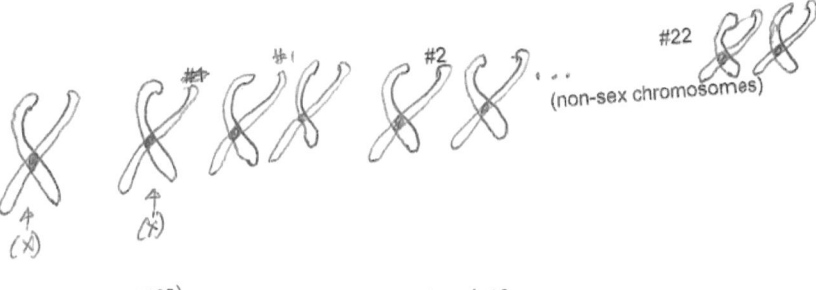

Female Human Karyotype

Female Human Karyotype

#1   #2   #3   ...   X   X

◁———— Non-sex chromosomes ———▷   Sex chromosomes

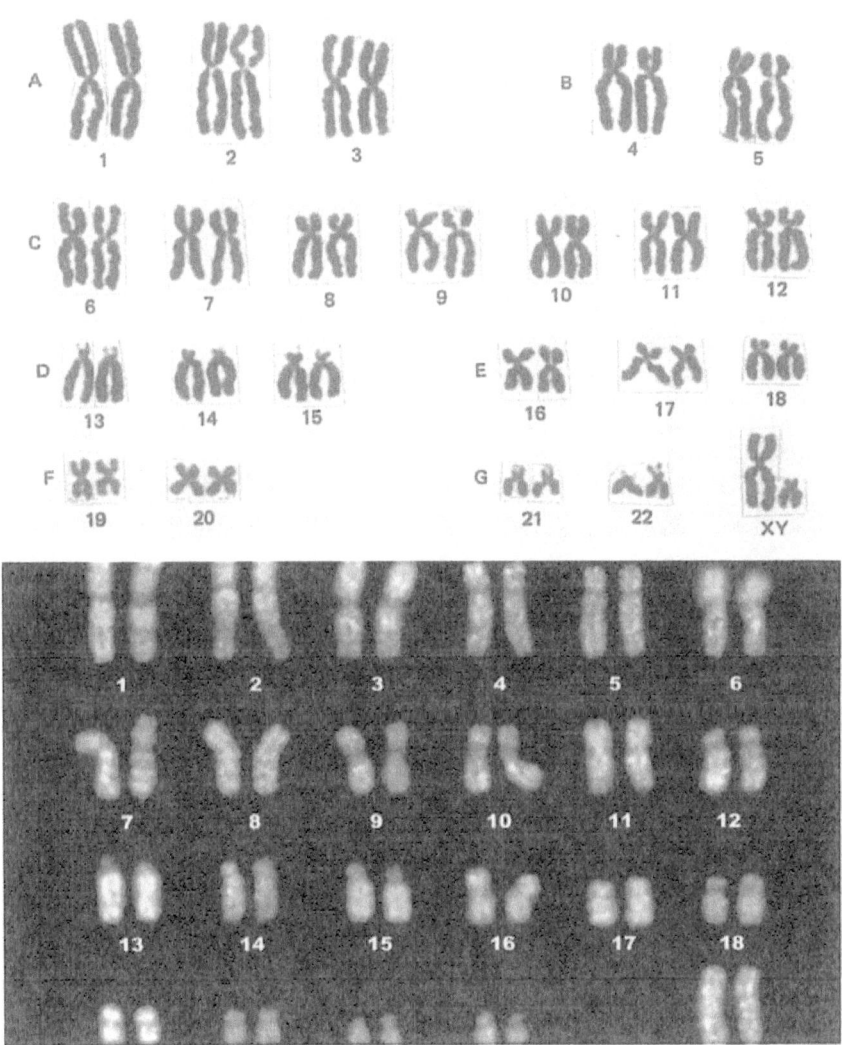

Karyotype

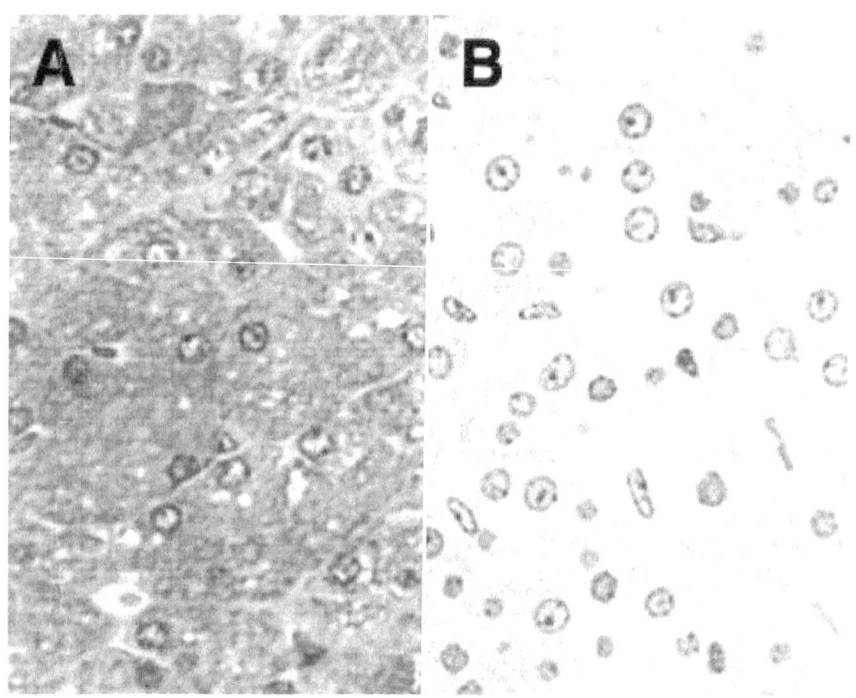

Cells with Giemsa Staining
(Organelles)
What do the organelles look like?

Then the daughter cells have to synthesize a replica chromatid.

Daughter cells synthesize new DNA.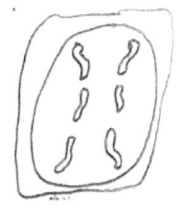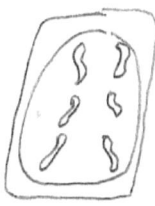

This is the synthesis or S-phase. It is during this phase that DNA is most susceptible to the effects of toxins, ionizing radiation, heat, viruses, and natural aging.

Finally, the adult cells have to grow again in what is known as the growth two or G2 phase.

The above can be summarized as:

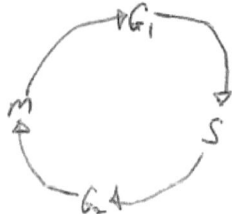

Cells that produce new cells in the skin, GI tract, bone marrow, and liver are stem cells. The others are committed cells and do not reproduce. In most tissues 2 percent of cells are stem cells.

The second form of cell division is meiosis, which only takes place in the organs of reproduction, the ovaries and testicles. These cells are referred to as gametes.

To a biologist an organism is an adult when it is old enough to reproduce.

Human ovaries contain 40,000 ova, and throughout a woman's reproductive life they secrete 400 of them. They somehow choose 1 percent of the ova that are of the highest quality. Fertility drugs such as Clomid take advantage of this. They act as false Follicle Stimulating Hormone (FSH) and cause more than 1 percent of the ova to be released. This can sometimes result in multiple deliveries at the same time. But it is nature's way of insuring that only the highest quality ova will be used to "propagate the species."

The principal difference between meiosis and mitosis is that in meiosis, chromosomes do not split into identical copies. The pair separates so that one daughter cell receives one set of alleles (such as blond hair, blue eyes), and the other gets the alternative allele (black hair and brown eyes).

During meiosis, chromosome pairs separate. Each chromosome has a different allele or genetic characteristic.

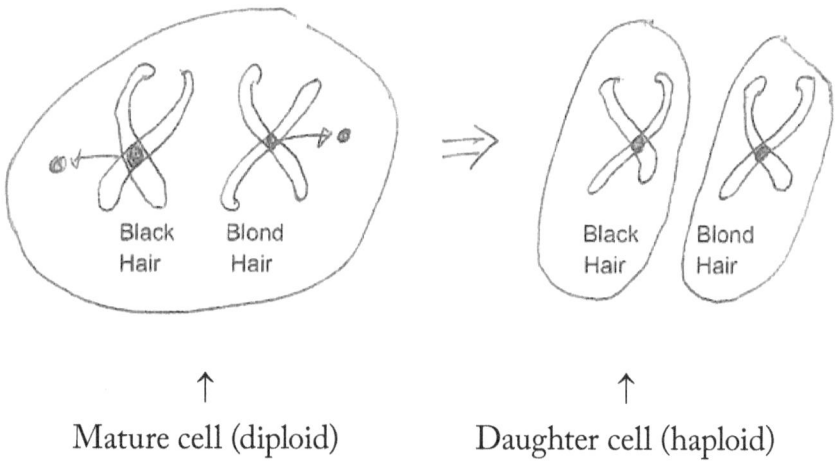

Mature cell (diploid)   Daughter cell (haploid)

With meiosis each daughter cell receives a random assortment of alleles or genetic characteristics. One might have blond hair and blue eyes and the other might have black hair and brown eyes. It represents a random assortment of genes and usually benefits the species.

As the chromosome pairs divide, the cell goes from a diploid state with pairs, to a haploid state with just one. And so gametes such as sperm and ova only have a haploid or half number of chromosomes. When two gametes join, for example a sperm fertilizes an ovum, a zygote forms that contains a diploid or double pair of chromosomes.

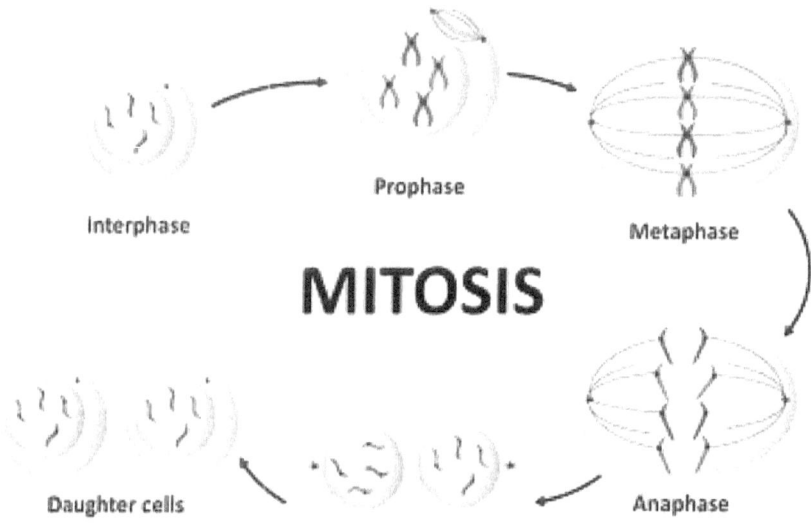

MITOSIS

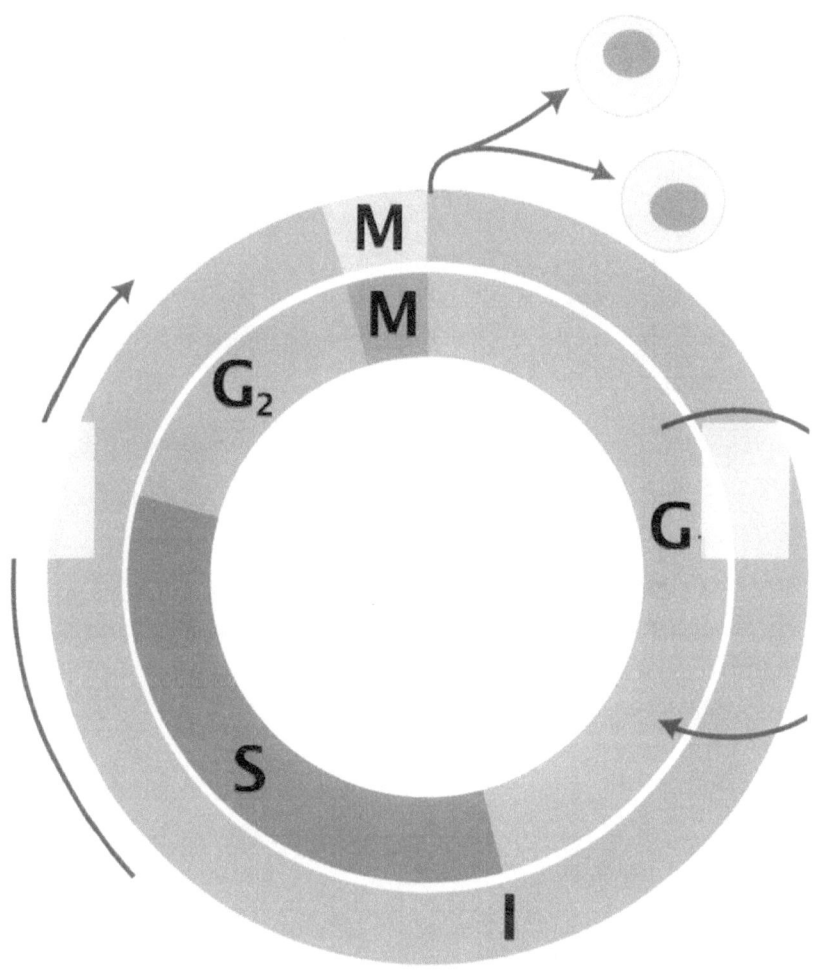

Cell Cycle

Chromosomes are held together by a centromere.

As we said, each chromosome comes as a pair. We have 23 such pairs, and each consists of an allele (an alternative form of the gene). As an example one allele might be for black hair and another for blond, one for brown eyes and the other for green eyes, etc. So the genome of a male would look like this:

 (Karyotype)

In its resting state, the genetic matter is diffusely distributed throughout the nucleus. When it goes through mitosis it passes through four phases:

- Prophase

  Metaphase
  Anaphase
  Telophase

In prophase the genes are diffusely distributed throughout the nucleus. In metaphase they become visible. In anaphase they line up along the cell's equator. In telophase, they divide and migrate to either end of the cell (sometimes known as the poles of the cell).

| Prophase | Metaphase | Anaphase | Telophase |

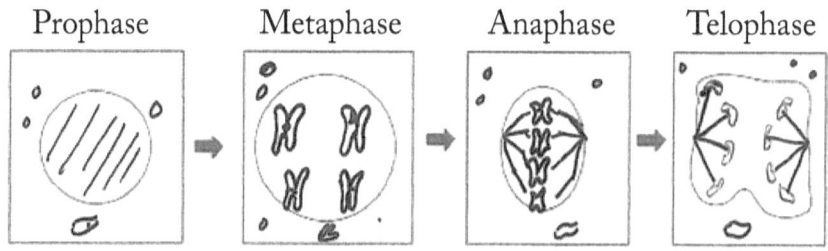

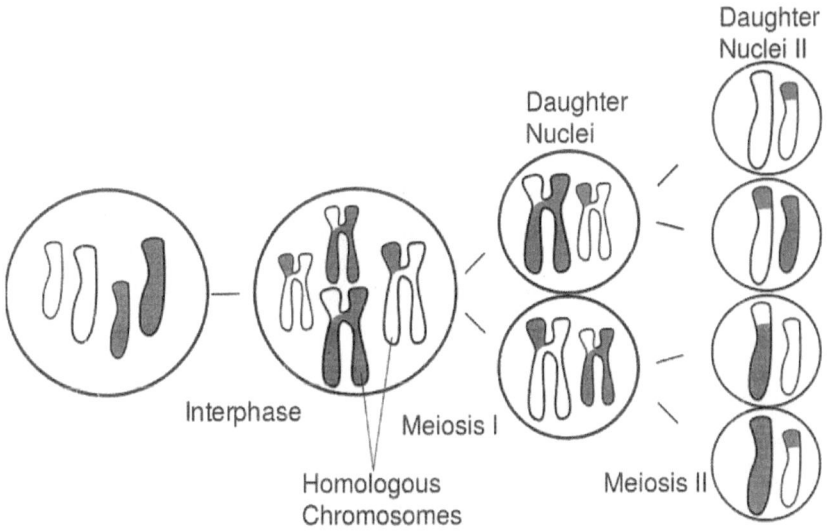

As chromosomes divide, pieces of one chromosome can become detached and reattach to another. This creates additional randomization of genetic characteristics.

Mitosis

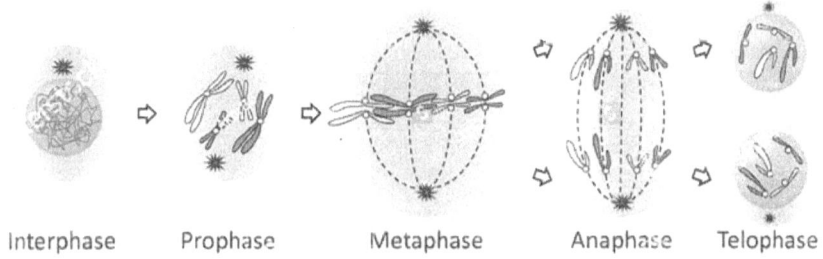

30)

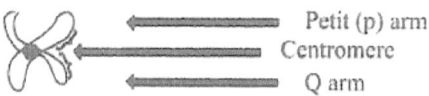
Petit (p) arm
Centromere
Q arm

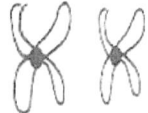
Chromosomes come in pairs. We have 23 pairs in our cells.

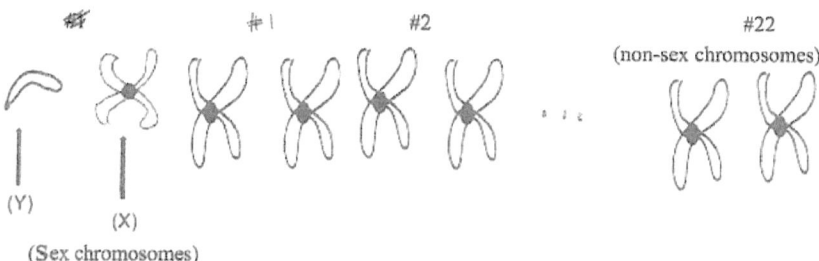

Male Human Karyotype

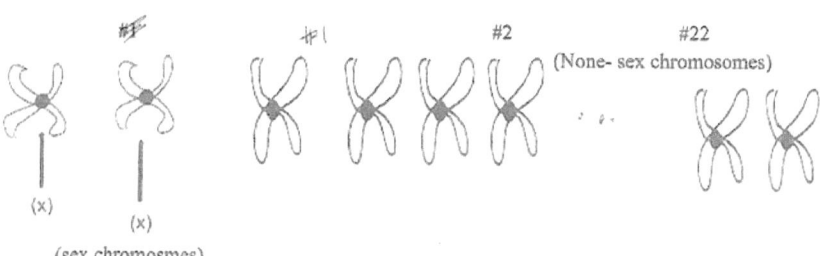

Female Human Karyotype

And it is the zygote that differentiates into an adult with a random assortment of genes.

Some of the genes are dominant and others are recessive. If an organism has a dominant and recessive allele of the same gene, it will present with characteristics of the dominant gene. Let's say that someone has a gene for black and blond hair. They would have black hair because black is dominant. We would say that they were heterozygous dominant. They would be heterozygous because they have two different genes. They would be dominant because they possess a dominant gene. If they had two genes for blond, they would be blond, and we would say that they were homozygous recessive. Homozygous because they possess two identical genes and recessive because they present with a recessive feature: blond.

The same principle would hold for:

- Brown eyes versus blue eyes

  Dark skin versus light skin

The science of genetics was developed by a monk named Gregor Mendel, who lived in a monastery in the second half of the nineteenth century. He worked with peas, and a branch of genetics (Mendelian genetics) was named in his honor.

To understand Mendelian genetics, we have to work with Punnett squares. Let us take our example of two people who have one gene for black hair and one for blond. Let B stand for black (the dominant gene) and b stand for blond (a recessive gene). Let us say one partner is heterozygous dominant (Bb) and the other is homozygous recessive (bb). What color hair would the children have?

	b	b (mother)
(father) B	Bb	Bb
b	bb	bb

Two children would have black hair and two would be blond. The children with black hair would be heterozygous dominant (Bb). The children with blond would be homozygous recessive (bb). The ones with black hair would be carriers. Even though they have black hair, they would carry a gene for blond.

This is actually a statistical probabilty. There is a 50% chance that two children would have black hair and two would have blond. There is no certanty that two would be blond and two would have black hair.

Suppose two carriers (Bb) were to marry. What would the children be?

		(mother)	
	B	b	
(father) B	BB	Bb	
b	Bb	bb	

Three of the children would have black hair and one would be blond. The first child would be homozygous dominant (BB). Two would be heterozygous dominant (Bb) and one would be homozygous recessive (bb). Again this is a statistical probability.

There would be a 75% chance that three of the children would have black hair and a 25% probability that one would be blond.

Supposing that a homozygous dominant (BB) were to marry a blond (bb).

	b	b
B	Bb	Bb
B	Bb	Bb

All of the children would have black hair and would be carriers (Bb).

These principles hold for many genetic characteristics, including undesirable ones such as Tay-Sachs and Huntington's chorea. Fortunately such genes are rare and are usually recessive. They would only manifest in the homozygous recessive form (i.e., bb). As long as people do not marry close relatives, their children should be free of the condition. And if they are in doubt, they may undergo genetic testing while the fetus is in utero. The question occasionally arises: if we sterilized individuals with genetic defects, could we eliminate them from the world? No. Within two generations they would be back as a result of genetic mutations.

Let us solve three more problems. What would happen if:

(1)
	B	b
B	BB	Bb
b	Bb	bb

(mother) = B, b (top)
(father) = B, b (left)

(2)
	B	b
b	Bb	bb
b	Bb	bb

(3)
	B	B
b	Bb	Bb
b	Bb	Bb

Addendum

1.	B	b
B	BB	Bb
b	Bb	bb

3.	B	B
b	Bb	Bb
b	Bb	Bb

2.	B	b
b	Bb	bb
b	Bb	bb

1. Both parents are heterozygous dominant (Bb):

    • One child would be homozygous dominant (BB).

    Two children would be heterozygous dominant (Bb), and they would be carriers.

    One child would be homozygous recessive (bb).

2. One parent is homozygous recessive (bb) and the other is heterozygous dominant (Bb):

    • Two of the children would be heterozygous dominant (Bb). They would also be carriers of the (b) gene. They would have black hair.

    Two would be homozygous recessive and would be blonds.

3. One parent is homozygous recessive (bb), and the other is homozygous dominant (BB).

    • All four children would be heterozygous dominant (Bb), and they would be carriers. They would have black hair.

Gregor Mendel

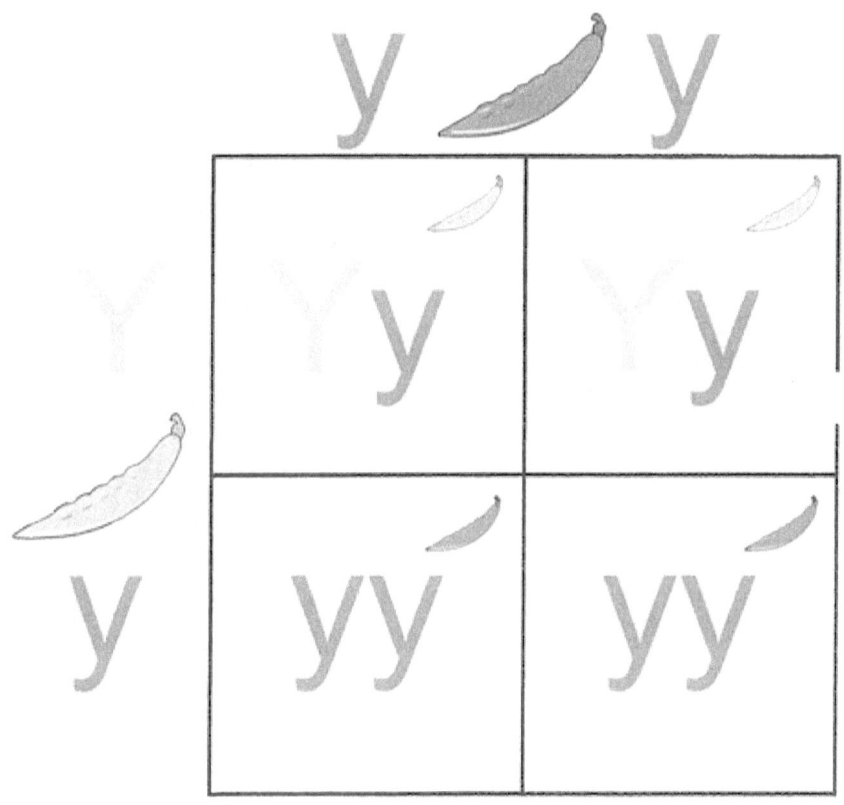

Punnett Squares

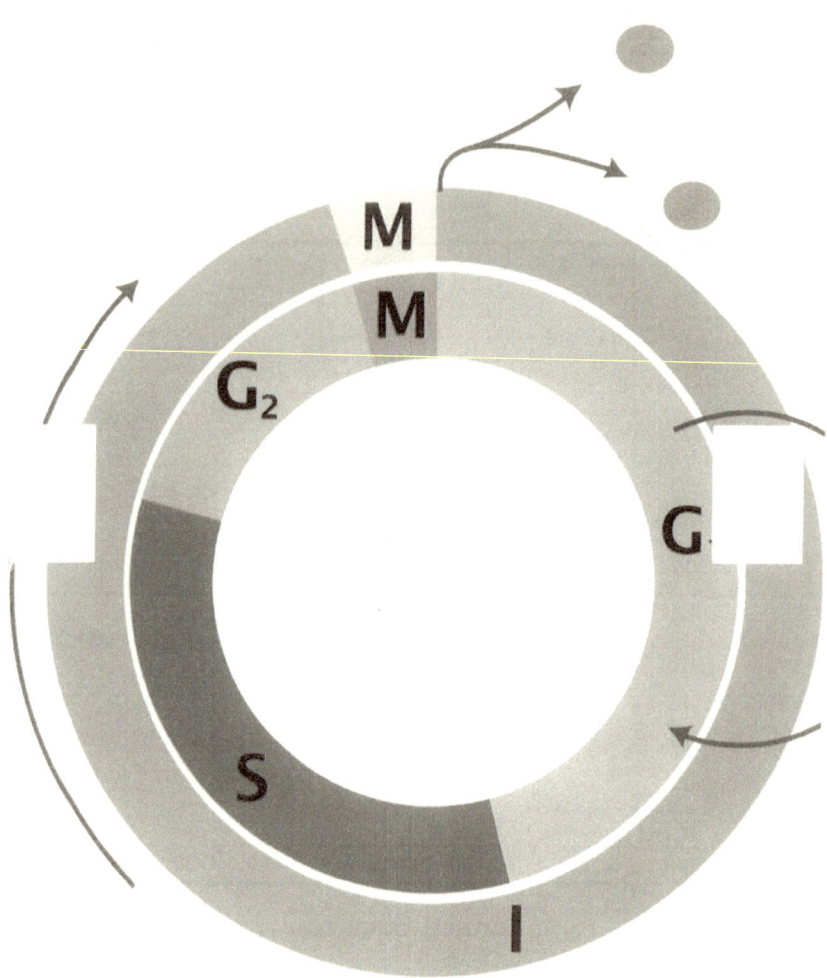

Cell Cycle

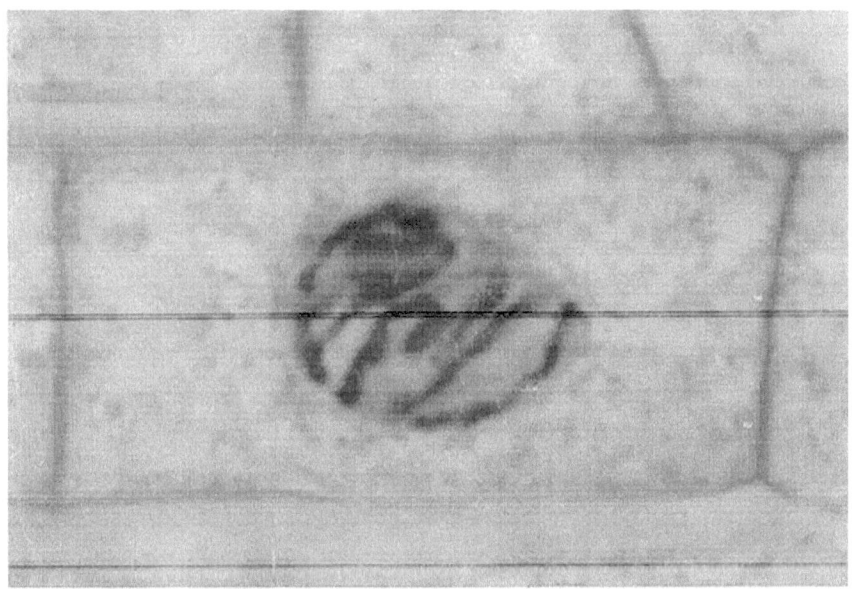

In prophase, the chromatin condenses into discrete chromosomes. The nuclear envelope breaks down and spindles form at opposite poles of the cell. Prophase (versus interphase) is the first true step of the mitotic process. During prophase, a number of important changes occur:

Prophase

- Chromatin fibers become coiled into chromosomes, with each chromosome having two chromatids joined at a centromere.

- The mitotic spindle, composed of microtubules and proteins, forms in the cytoplasm.

- The two pairs of centrioles (formed from the replication of one pair in Interphase) move away from one another toward opposite ends of the cell due to the lengthening of the microtubules that form between them.

- Polar fibers, which are microtubules that make up the spindle fibers, reach from each cell pole to the cell's equator.

- Kinetochores, which are specialized regions in the centromeres of chromosomes, attach to a type of microtubule called kinetochore fibers.

- The kinetochore fibers "interact" with the spindle polar fibers connecting the kinetochores to the polar fibers.

- The chromosomes begin to migrate toward the cell center.

Metaphase

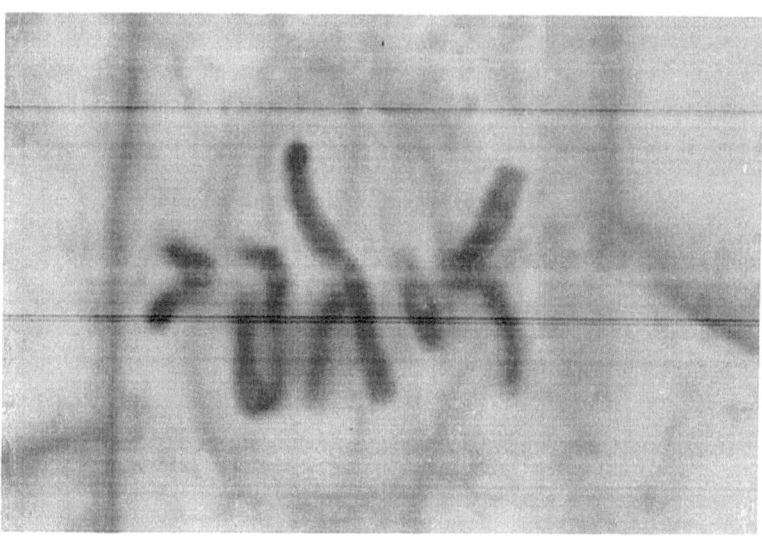

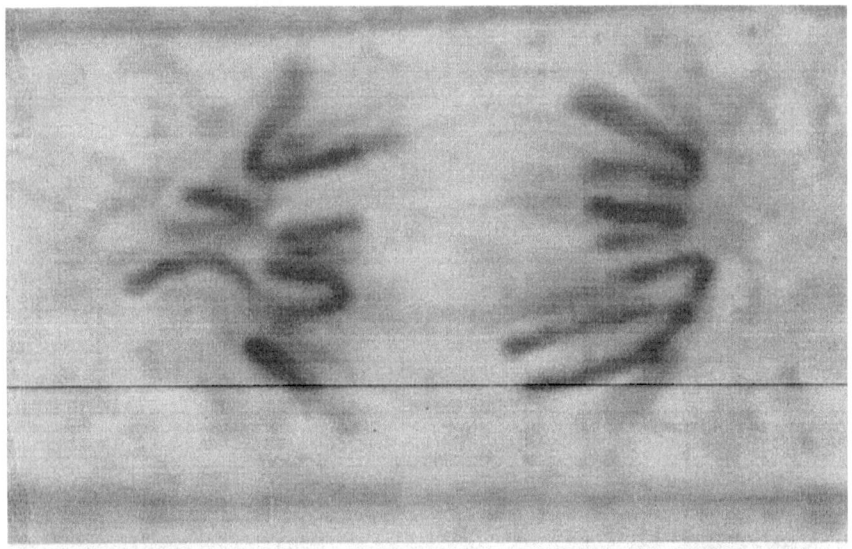

In anaphase, the paired chromosomes (<u>sister chromatids</u>) separate and begin moving to opposite ends (poles) of the cell. Spindle fibers not connected to chromatids lengthen and elongate the cell. At the end of anaphase, each pole contains a complete compilation of chromosomes. During anaphase, the following key changes occur:

Anaphase

- The paired centromeres in each distinct chromosome begin to move apart.

- Once the paired sister chromatids separate from one another, each is considered a "full" chromosome. They are referred to as daughter chromosomes.

- Through the spindle apparatus, the daughter chromosomes move to the poles at opposite ends of the cell.

- The daughter chromosomes migrate centromere first and the kinetochore fibers become shorter as the chromosomes near a pole.

- In preparation for telophase, the two cell poles also move further apart during the course of anaphase. At the end of anaphase, each pole contains a complete compilation of chromosomes.

Telophase

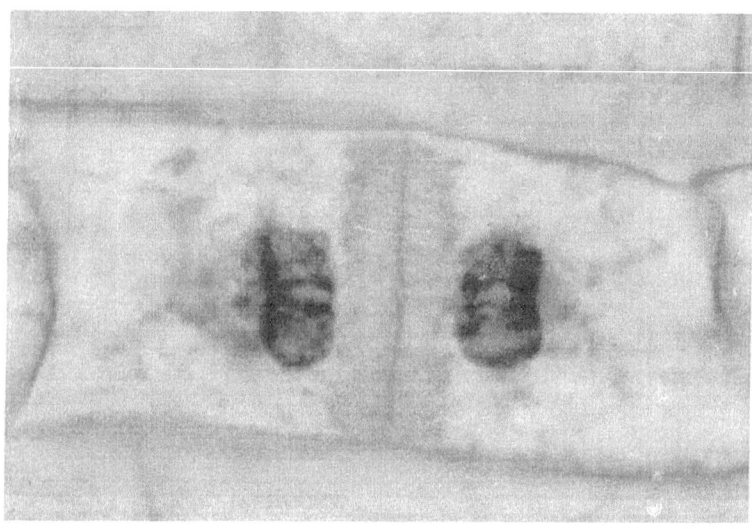

The Stages of Mitosis and Cell Division

Mitosis is the phase of the cell cycle where chromosomes in the nucleus are evenly divided between two cells. When the cell division process is complete, two daughter cells with identical genetic material are produced.

Interphase

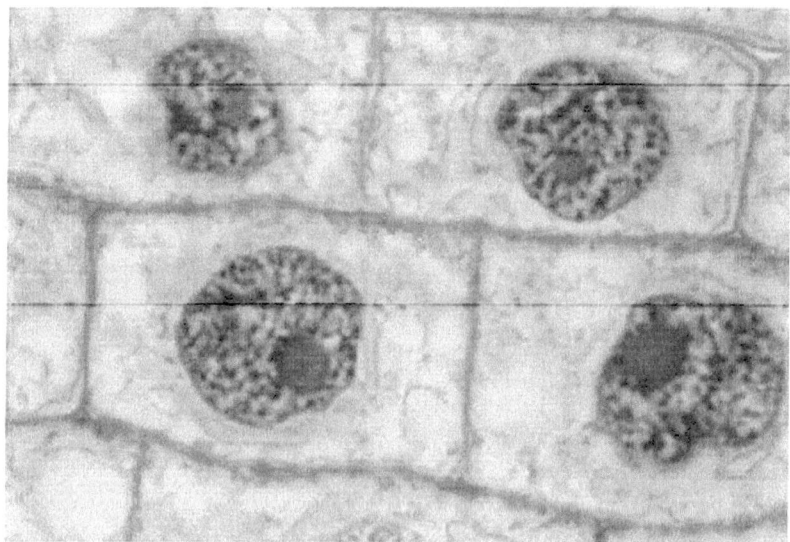

Meiosis I in Females

Prophase I
chromosomes begin to condense

recombinant chromosomes

Metaphase I
spindle fibers attach to chromosomes
chromosomes line up in center of cell

Anaphase I
chromosomes start to move to opposite
ends of cell as spindle fibers shorten

Telophase I
chromosomes reach opposite ends
nuclear membrane forms

Cytokinesis
cell division occurs

Meiosis

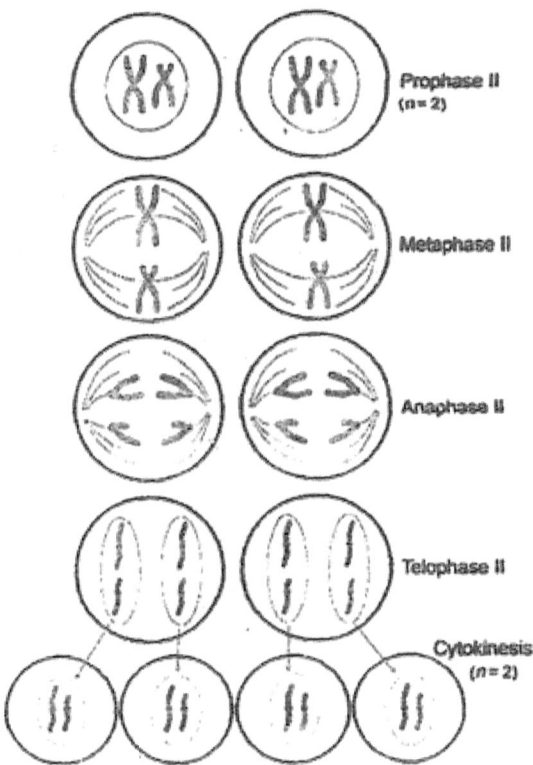

Prophase II
(n = 2)

Metaphase II

Anaphase II

Telophase II

Cytokinesis
(n = 2)

The daughter cells are . Jan 21, · Cell Division: Mitosis and Meiosis.

Meiosis I

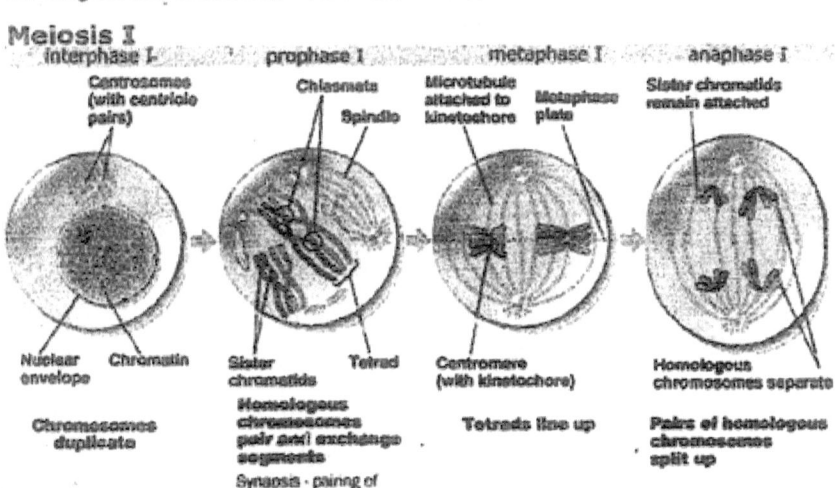

interphase I   prophase I   metaphase I   anaphase I

Centrosomes   Chiasmata   Microtubule   Sister chromatids
(with centriole   attached to   Metaphase   remain attached
pairs)   Spindle   kinetochore   plate

Nuclear   Chromatin   Sister   Tetrad   Centromere   Homologous
envelope   chromatids   (with kinetochore) chromosomes separate

Chromosomes   Homologous   Tetrads line up   Pairs of homologous
duplicate   chromosomes   chromosomes
                         pair and exchange   split up
                         segments
                         Synapsa · pairing of
                         homologs to form

Not all genes work on the principle of Mendelian genetics. We need to define:

Additive genes

Incomplete penetrance

Mosaics

Additive genes are ones where the more you have the greater the characteristic. Being tall is a good example. Chances are that the more "tall" genes you have, the taller you will be.

Incomplete penetrance means that two genes express themselves equally. Suppose we cross a red flower with a white and get a pink flower. That would be incomplete penetrance also known as codominance.

A mosaic is a phenomenon in which alternative genes are expressed equally in different regions of the same body. Examples include:

* Calico cats

  Zebra
  Leopards
  Tigers
  Flowers with red and white stripes

One of the most interesting aspects of genetics is the phenomenon of promoter and suppressor genes. These are genes that turn other genes on or off. This protects us against cancer by turning oncogenes off (genes that cause tumors). They can also turn growth genes on, which leads to "growth spurts" in children.

We should conclude by defining two new terms: genotype and phenotype. An organism's genetic makeup is its genotype. Phenotype is determined by the way that genes interact with the environment. Let us say for example, that someone has genes for a tall stature but consumes a poor diet as a child. When they reach adulthood they might be below average height. So it is a matter of genetic makeup and the environment—genotype and phenotype.

One of the most important reasons for learning genetics is to understand evolution. In his famous book *On the Origin of Species* published in 1859, Charles Darwin theorized that life is a struggle to survive, to find sufficient food and shelter, and coined the term "survival of the fittest." He observed that no two species could occupy the same niche in an ecosystem. They could not eat the same food in the same place at the same time. If they did, one of them had to go, and it was survival of the fittest. The organism whose genes made it most suitable would predominate. The others would go extinct. This is how life evolved on Earth.

There are two periods in Earth's history that are remarkable. Ice ages last for variable periods. The longest occurred 720 million years ago and lasted for 85 million years. The other was the Devonian which occurred 380 million years ago.

Ice ages are caused by:

- The Earth shifting away from the sun.

  The average temperature dropping by 16 degrees Fahrenheit.

  Earth being covered by glaciers.

Charles Darwin

The *Beagle*, the ship that Charles Darwin used to
explore South America and the Galapagos Islands.

Only the hardiest forms of life could survive an ice age, and it is thought that only the "smartest" humans could have survived. Multicellular and single-celled organisms once lived under ice for 85 million years!

Addendum

Early humans discovered fire, and learned to make clothes, weapons, tools, and primitive houses. They also domesticated the dog. Neanderthals did not and could not survive the ice ages.

The Devonian was a period of "massive droughts" in which marine animals and plants found themselves on dry land. The overwhelming majority perished. A very small number were able to use ambient air, such as today's catfish and prawn. An even smaller number also developed bony fins that allowed them to ambulate (like today's catfish). Some of these fish developed a primitive lung (like the catfish). This gave them four advantages:

- They could breathe.

  They could walk.

  They had an abundant supply of food (the other fish that had been stranded).

  They had no predators.

The Devonian Period occurred 380 million years ago and lasted 60 million years. No one knows what caused it. The leading theory is that it was plate tectonics; the idea that the earth's crust can be divided into plates that are the size of continents. These plates float on liquid rock called magma, deep within the crust. Something caused them to rise, and areas that had once been on the bottom of the ocean suddenly became dry land.

We should explain something about reproductive biology. If you placed a male and female mouse in a field with abundant food and no predators, you would have a million mice in a year. After two years you would have a million million (a thousand billion). So even two organisms with lungs and bony fins could have populated the Earth in a few decades.

So the early land dwellers flourished. Some went on to become land animals, and others went back to the ocean. Those who returned to the sea retained their bony skeletons and their lungs became air organs. These organs allowed them to change depth; hence the Osteichthyes came into existence (fish with bony skeletons).

Land organisms and those that lived in the ocean continued to evolve. Evolution can take place faster than is commonly believed. Darwin raised finches in an aviary. Within a few generations they developed shorter wingspans to cope with their smaller habitat. Today there are 6 million different life forms on earth.

So how do we classify life forms? A philosopher named Carl Linnaeus devised a seven-part system. He placed all living things into one of seven categories.

Carl Linnaeus
1707–1778

First to devise a classification system for living organisms:

• Kingdom

Phylum
Class
Order
Family
Genus
Species

Classification of Living Organisms

Let us talk about how biologists classify living organisms. They typically use the mnemonic "kids prefer candy over fancy green salads."

- Kingdom

  Phylum
  Class
  Order
  Family
  Genus
  Species

Kingdom: Is an organism an animal, a plant, or a microbe? Does it have some form of locomotion, even for only part of its life cycle? If so. it is an animal. Does it stay in one place throughout its life? If so, it is a plant. Do we need a microscope to see it? If so it is a microbe.

Some plants have muscle tissue, as attested to by the fact that there are 750 species of carnivorous plants: they use modified muscles to catch prey. But they do not get up and move around.

Phylum: Once having decided whether an organism is a plant, an animal, or a microbe, one has to ask do they live in salt water, freshwater, or on land? One also has to decide whether their bodies have radial symmetry (round like a pie), biradial symmetry (an incomplete circle), or bilateral symmetry, in which one side is a mirror image of the other.

Class: is a particular life form a:

- Helminth: a worm

  Molluscum: soft-bodied
  Crustacean: with an external skeleton
  Vertebrate: with an internal skeleton and a spine

One of the broadest groupings is that of the soft-bodied animals, molluscum, which includes:

> Snails
> Slugs
> Shellfish
> Octopi

Crustaceans include:

> Lobsters
> Crabs
> Echinoderms (prickly-skinned ones: sand dollars, sea urchins)
> Squid
> Insects
> Nautilus

Order:

Then we have to ask several questions. If it is an animal, is it warm-blooded or cold-blooded? Is its body temperature the same as the environment (cold-blooded), or is it a homeotherm (one that generates heat)?

Cold-blooded organisms include:

- Amphibians (living mostly in water)

Reptiles

Warm-blooded organisms include:

- Birds

  Marsupials
  Mammals

We now have to define the basic life forms: What is an amphibian? An animal that spends most of its time in water. There are two reasons for this. One, they have thin integuments (skins) and cannot retain water; and two, they have a primitive renal system (kidneys) and cannot retain water.

Their skins can exchange oxygen and carbon dioxide with the surrounding water. Examples:

- Frogs

  Toads
  Salamanders
  Newts

What is a reptile? A cold-blooded vertebrate that lives on land and in water. In general they:

- Have leathery skin

  Grow as long as they live
  Are usually carnivorous
  Can usually grow back lost appendages

They include:

- Snakes

  Lizards
  Crocodiles
  Alligators
  Tortoises
  Turtles

Addendum

(Lizards)

There are 5,000 varieties of lizards. The largest is the Komodo of the southern Pacific. They are also called "dragon" lizards and are about the size of crocodiles.

They feed upon medium-sized mammals, such as goats and deer, and snakes. They have a tough hide and operate off two emotions. If they are hungry they lose their fear and leave their caves. If they are not hungry, they are afraid and stay within the caves.

Unlike mammals, reptiles do not have natural antiseptics in their saliva. They build up a cocktail of bacteria in their saliva from decaying meat products. When they bite, they inject bacteria, and the prey dies of septicemia (infection of the blood) within hours to days. The Komodo will follow it until it collapses.

This is an effective toxin but not a true venom (which is a modified saliva). The only lizard on Earth that is truly venomous is the Gila monster of southwest United States and Mexico. They store venom in their lower jaw and inject it through a pair of hollow fangs.

Other lizards (to name a few) include:

- The chameleon, which can change color

  The horned toad

  Small lizards of the North American continent

Most consume insects and pose little threat to humans.

What is a fish? A swimming vertebrate with gills, who lives in saltwater and/or fresh water. They typically lay eggs and have:

- Scales

  Tails

  A two-chambered heart

  Dorsal and ventral fins

  They include Chondrichthyes (those with cartilaginous skeletons)

    Sharks
    Sting rays
    Bat rays
    Manta rays

These fish do not possess inflatable organs and sink if they do not swim.

Osteichthyes have bony skeletons and inflatable organs which allow them to change depth. They include:

- Carp

  Salmon
  Tuna
  Eel
  Bass

What is a bird? A warm-blooded vertebrate that can fly or could fly in the past. Penguins, ostriches, emus, and kiwis are flightless birds, but they could fly millions of years ago.

Birds:

- Have large lungs

  Possess large muscle mass

Konrad Lorenz worked out the theory of "imprinting." This takes place in several species of birds. When chicks are born (hatch), they adopt the first living thing they see as "mother." They follow it everywhere until maturity. Any living thing they see after that is an enemy to be avoided.

Birds:

- Have hollow bones
- Have air-filled feathers
- Have a beak
- Have bipedal locomotion (they walk on two legs).
- Many can run fast. Ostriches have been clocked at 45 mph.
- Lay eggs (they possess a tubular organ called a cloaca that produces eggs)
- Do not have mammary glands
- Are good parents.
- Have good eyesight.

Birds that fly do so by moving their wings as if they were rowing a boat, thrusting down and back, thereby giving themselves lift and forward motion. Birds learn to fly from their parents.

Marsupials are warm-blooded vertebrates that produce live birth and carry them in a pouch until they are mature enough to live independently. Examples include:

- Kangaroos

  Opossums
  Platypuses
  Anteaters

Mammals are warm-blooded vertebrates that give birth to live young and have mammary glands with which to nourish their young. They:

- Are homeothermic.

  Have ovaries, uterus, and a vagina or a penis.

Parent their young in early life.

Having two or more mammary glands with which to feed newborns until they can eat and are weaned.

Family:

So what family does an organism belong to? Let us say they are mammals. They may belong to any number of families:

- Weasels (minks, ferrets, badgers, and wolverines).

  Bears (related to raccoons).

  Felines (sand cats, feral cats, margays, ocelots, bobcats, lynx, leopards, cheetahs, jaguars, mountain lions, lions and tigers)

Cats can be divided into two groups. There are the Pantheras and the non-Pantheras. Large cats belong to the group Panthera. They include:

> cougars
> pumas
> leopards
> cheaters
> jawars
> lions
> tigers

non-Pantheras include:

> house cats
> sand cats
> bob cats

margays

ocelots

lynx

Canines (dogs, wolves, foxes, coyotes, jackals, hyenas, and dingoes)

Equestrians (horses, zebra, ponies, and donkeys)

Antelopes

Swine (pigs, boars)

Rodents (mice, gophers, rats, squirrels, chipmunks, hamsters, guinea pigs, and beavers)

Pachyderms (the thick-skinned ones: elephants, rhinoceroses, and hippopotamuses)

Animals with split hooves (deer, moose, elk, gazelle, goat, sheep)

Relatives of Primates (lemurs, monkeys, chimpanzees, gibbons, orangutans, gorillas, and *Homo sapiens*)

Ninety-nine percent of a gibbon's genes are identical to those of a human. The one percent that is different made it possible for us to build the pyramids, write the Magna Carta and plays of Shakespeare, compose the symphonies of Beethoven, split the atom, cure cancer, and travel to the planets.

Lepidorae (rabbits, hares)

Bovines (cow, steer, ox, buffalo)

There are eight species of bear throughout the world. One species is purely carnivorous (eats only meat), the polar bear, and one is purely herbivorous (eats only plants), the panda. The others are omnivores and eat both meat and plant materials (20 percent meat and 80 percent plants).

Bears have sharp teeth. Most also have nonretractable claws that can be used as weapons or for digging and climbing trees.

They go into a prolonged sleep for one to four months during the winter. It is not true hibernation and their cubs are often born during this time. They begin the first few months of their life in the mother's den.

Their front legs are shorter than their hind legs but they can still run at high speeds (35 to 50 mph). Their nearest relative is the raccoon.

Let us talk about cats. Cats have:

- Good eyesight (six times better than a human)

  Good hearing
  Good sense of smell
  Sharp claws and teeth
  Flexible backs
  Large muscle mass
  The ability to jump high and to run at high speeds

Cats are purely carnivorous and learn to hunt from their mothers. When born she will nurse them for six months. Then they will accompany her for a year and a half while she teaches them to hunt. Then she will wander off. Brothers and sisters will stay together for another year, but eventually they too will separate (except for lions). This reduces competition for food and mates.

Cats sleep sixteen to twenty hours a day and become active at night. This allows them to conserve energy, stay cool, avoid enemies, and hunt at a time when they can see the prey, but the prey cannot see them.

Large cats need 400 square miles of territory to find sufficient food, shelter, and mates.

Supposing it is a dog? Dogs:

- Usually live in packs.

  Hunt together.
  Are omnivores.
  Have sharp teeth.
  Have nonretractable claws that are good for digging.

Addendum

Genus includes organisms that are closely related and could interbreed. But their progeny would be sterile. Examples include:

- Mules

  Ligers
  Seedless fruits
  Annual flowers
  Cherry trees that do not produce cherries

This is because members of the same genus have different numbers of chromosomes.

All members of the same species can produce fertile offspring because they have the same number of genes, e.g.:

- All house cats

  All dogs
  All horses
  Perennial flowers (that bloom every year)

Let us talk about another class that includes a wide variety of organisms: Mollusca.

Squid

Octopus

Nautilus

Many mollusks, such as slugs, snails, and abalone are pseudopods, a name meaning "false footed." It is a suction-cuplike appendage that makes locomotion possible.

The valve mollusks such as abalone, mussels, clams, and oysters have filtration systems that they use to obtain nutrition. An oyster can filter 100 gallons a day. When eating them, one has to be careful that they did not come from waters that contain the hepatitis virus.

One of the most interesting mollusks is the octopus. They are the most intelligent of the invertebrates and have three hearts: one that pumps blood to the gills and two that pump to the rest of the body. They have biradial symmetry, a liquid propulsion system, reversible chromatophores which allow them to change color, dispensable ink, blue blood, and a sharp beak. They are very pliable and can squeeze under the bottom of a door. Some, such as the ringed octopus, are poisonous. Two-thirds of their brain is in their arms, and they can regrow lost limbs.

Anatomy

Let us shift gears and talk about a vast topic that we can only touch upon: anatomy and physiology. We will talk about the human body.

There are two ways to study anatomy and physiology. We could take the systemic approach and talk about each organ system. Or we could use regional systems: first learn all there is to know about the hand, then move on to the arm, and so on.

The systemic approach is usually easier. Let us talk about:

- The cardiovascular system

    The pulmonary system
    The digestive system

Leonardo da Vinci

Created the first text on human anatomy

In the late 1400s and early 1500s

- The central nervous system

  The musculoskeletal system
  The hematogenous system
  The renal system
  The skin

Anatomy is a study of the organs and their position relative to each other. Physiology is the way that the different organs work.

The cardiovascular system begins with the heart. The heart is the size of an adult fist and beasts 60 to 100 times a minute. It consists of four chambers, four valves, and an electrical system. Impulses delivered by the vagus and sympathetic nervous systems enter the heart through the sinoatrial node (SA node) at the top of the right atrium. Then they travel through nerve fibers that run between the two atria. They enter the atrioventricular node (AV node) at the top of the two ventricles. From there they progress through the bundle of His (specialized nerve fibers that run along the interventricular wall between the ventricles). Once they reach the apex (at the bottom of the heart) they change direction and course through the Purkinje fibers, which run along the outer walls of the two ventricles.

As neurological impulses travel through the heart, they trigger a wave of contraction.

Blood returns to the heart by way of the superior and inferior venae cavae. The superior vena cava drains the upper half of the body, and the inferior vena cava drains the lower half. The blood pressures in the inferior vena cava and superior vena cava are the diastolic (lower number). This is the pressure that fills the heart.

When blood enters the heart it fills the right atrium. As the atrium contracts, it forces blood through the tricuspid valve (with three leaves) into the right ventricle. From there blood is pumped through the pulmonic valve into the pulmonary artery (the only artery to carry deoxygenated blood).

The pulmonary artery takes it to the lungs. Lung conditions that cause pulmonary hypertension can also cause cor pulmonale (enlargement of the right ventricle).

From the lungs blood returns by the pulmonary vein (the only vein to carry oxygenated blood). Then it enters the left atrium and passes through the mitral valve (two-leaved). Then it enters the left ventricle, which pumps it through the aortic valve. From there it enters the aortic arch, which takes it to the body.

When discussing the tricuspid and mitral valves remember the acronym 'try before you buy.' Blood passes through the tricuspic first. Then the mitral which is a two leaf valve.

The pressure of the blood as it leaves the heart is the systolic (the upper number). So let us say that the blood pressure is 120/80. This means that blood leaves the heart under 120 bars of mercury and enters it with 80 bars.

Once blood leaves the heart, it enters the aortic arch. The orifices of the coronary arteries are located on the aortic wall just distal to the aortic valve. They provide for the heart itself.

The left and right common carotids branch off the aortic arch and carry blood to the upper body. On the left the brachial artery branches off the aortic arch and supplies the left arm. The right brachial branches off the right common carotid and supplies the right arm. So blood pressure is usually taken from the left which is thought to be more accurate.

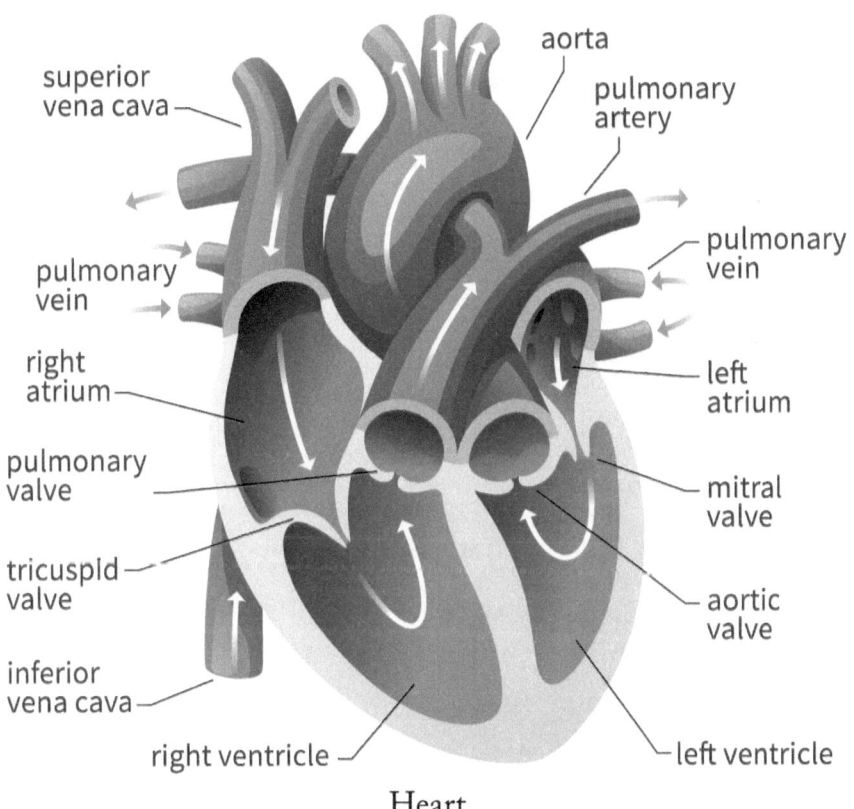

superior
vena cava

aorta

pulmonary
artery

pulmonary
vein

pulmonary
vein

right
atrium

left
atrium

pulmonary
valve

mitral
valve

tricuspld
valve

aortic
valve

inferior
vena cava

right ventricle

left ventricle

Heart

The common carotid then branches into the left and right internal carotids and the external carotids. The left and right external carotids supply the anterior (front) parts of the head.

The left and right internal carotids join the circle of Willis (which encircles the brain) and provide it with two-thirds of its blood.

The aortic arch flanks the thoracic spine (backbone behind the chest) as it descends into the upper body: veins, arteries, and nerves come off the spine, the inferior vena cava, and the aorta and course along the under surface of the ribs, forming the acronym *VAN*: vein, artery, and nerve. The veins are near the top of the ribs. The arteries are in the middle, and the nerves are at the bottom.

Once the aorta drops below the diaphragm, it gives rise to several branches that supply the abdominal organs. The left and right renal arteries supply the kidneys and the adrenal glands.

Then the gastrosplenic artery supplies the stomach and the spleen.

Then the superior and inferior mesenteric arteries join the intestinal artery to provide for the intestines.

Eight arteries provide for:

- Adrenal glands

  Kidneys
  Stomach
  Spleen
  Intestines

  Liver (hepatic arteries branch off the celiac artery; a branch of the hepatic artery supplies the gallbladder)

When the aorta reaches the pelvis it branches into the left and right iliacs, which supply the pelvic organs:

- Uterus

  Vagina
  Fallopian tubes
  Bladder
  Prostate
  Penis
  Rectum

Once the iliacs drop below the inguinal canals and enter the legs, they become the femoral arteries. They course with the neurovascular bundle, which is close to the femur. Once the arteries go below the knees, they become the popliteal arteries, which supply the lower legs and drain into the plantar arteries, which supply the feet.

When blood returns, it does so through veins, which course with the arteries and have much the same name: the popliteal veins, femoral veins, and iliac veins.

Human Circulatory System

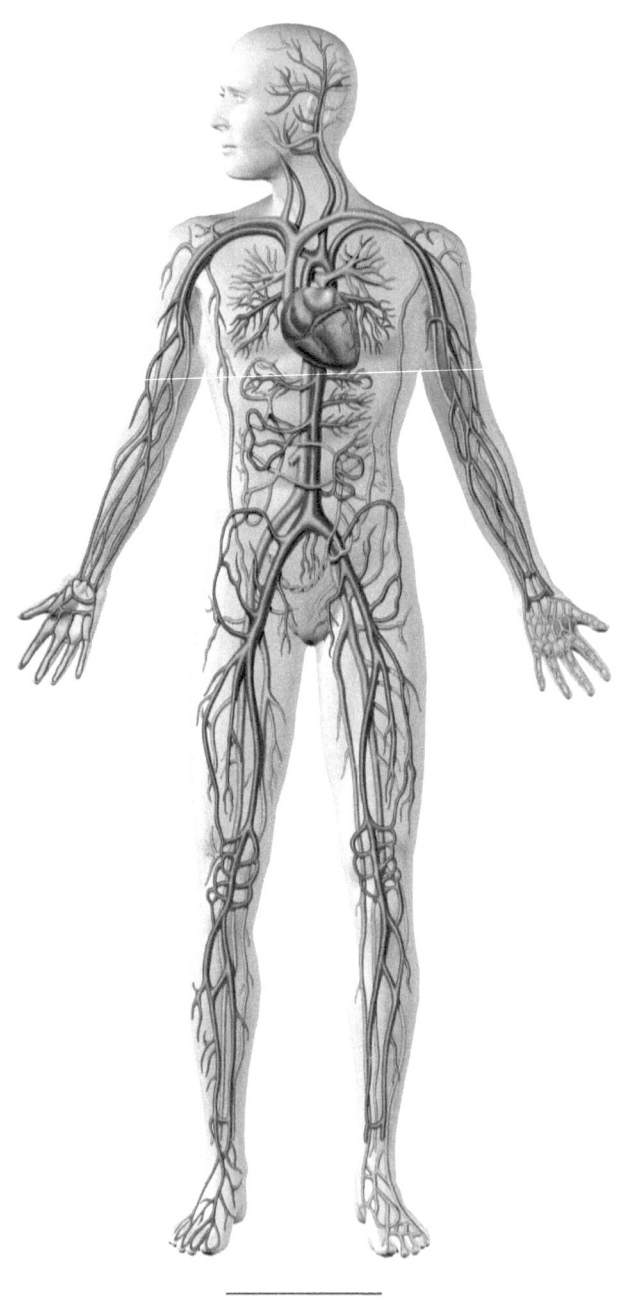

The vessel that returns blood from the lower body starting at the pelvic brim is the inferior vena cava. The veins that return blood from the head are the left and right jugular, and the main vessel is the superior vena cava.

Let us follow two drops of blood as they travel through the cardiovascular system. One drop will enter the heart from the superior vena cava, and the other from the inferior vena cava.

First they enter the right atrium.

Then they pass through the three-leafed valve called the tricuspid. Then they will enter the right ventricle. From there they will be pumped through the semilunar valve (which means 'partial moon') into the pulmonary artery (the only artery to carry deoxygenated blood). From there they will go to the lungs, acquire oxygen and return to the heart in the pulmonary vein; the only vein to carry oxygenated blood.

The two drops enter the left atrium and are pumped through a two-leafed valve called the mitral. Now they enter the left ventricle, which will pump them through the aortic valve and allow them to leave the heart and enter the aortic arch.

The two drops go to the heart itself via the coronary arteries, the orifices to which are on the wall of the aorta just as it leaves the heart.

Then the two drops continue down the aortic arch until they reach the carotid arteries. Some of the blood goes up the right carotid until it reaches the right brachial artery where some goes to the right arm. The rest would go to the bifurcation of the common carotid, where they give rise to external and internal carotids. The blood that enters the external carotid goes to the anterior (front part of the head). The blood that enters the internal carotid goes to the circle of Willis where it supplies the brain.

Further down the aorta some of the blood enters the left brachial artery to supply the left arm. As the blood continues down the aorta, it supplies the ribs and associated structures, and when it reaches the diaphragm it supplies the abdominal organs: liver, adrenals (whose artery branches off the renal artery), kidneys, spleen, gallbladder (branching off the hepatic artery), stomach, and intestines.

Remember that the aorta courses with the thoracic and lumbar spines (the backbone of the chest and abdomen). Small blood vessels extend from the aorta and supply the vertebrae (small bones in the back and neck). Another set of vessels supplies the spinal cord. They feed into three spinal arteries that course with the cord: two posterior spinal arteries and one anterior spinal artery. The three arteries also supply the posterior part of the brain with one-third of its blood.

Then the two drops enter the iliac arteries, which supply the pelvis.

In women:

- Ovaries

  Fallopian tubes
  Uterus
  Vagina
  Sigmoid colon
  Bladder

In men:

- Bladder

  Prostate
  Testicles
  Penis
  Sigmoid colon

Then they enter the femoral arteries and supply the upper legs (thighs). Once below the knees, they enter the popliteal arteries and supply the lower legs. In the feet they enter two planar arches with branches to the toes.

Veins accompany the arteries and usually have the same name. The exceptions are listed above.

At this point you might be wondering two things;

- What are the differences between veins and arteries?

  How does skin get its blood?

Most arteries contain a layer of smooth muscle (involuntary muscle). It is referred to as the musculares and serves three functions:

- It maintains blood pressure.

  It helps propel the blood through the body.

  It seals the blood vessels in the advent of a severe laceration.

A few exceptions include arteries of the brain, which do not have a muscularis and cannot stop bleeding if damaged.

Veins do not contain a muscular layer. They do have a one-way valve that only permits blood to flow toward the heart—unless they break down, as with varicose veins.

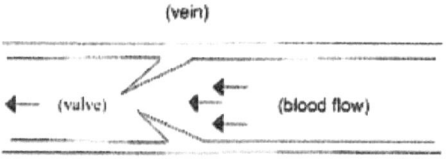

So how does skin receive blood? The neurological innervation of the skin comes from dermatomes. Circular patches of the skin that extend from the spinal cord.

Arteries and veins accompany the dermatomes and supply the skin.

Now let us switch gears and think about the heart in another way: in terms of electrophysiology. We need to review the heart's electrical system and talk about one of the most frequently ordered tests in medicine, the EKG.

The heart is driven by two nervous systems: the sympathetic and parasympathetic. The sympathetic speeds the heart up and increases blood pressure. The parasympathetic does the opposite.

The parasympathetic is the vagus nerve, which is the tenth cranial nerve and originates in the brain: (X cranial nuclei). It secretes a neurotransmitter called acetylcholine. Sympathetic stimulation comes from the sympathetic chain ganglion, which runs along the sides of the thoracic and lumbar spines. It uses norepinephrine and epinephrine as its neurotransmitters.

To make the heart run faster, the sympathetic chain ganglia stimulates alpha and beta receptors. If the heart were to lose its connection with the sympathetic and vagus nervous systems, or if the connection between the SA node and the AV node were lost, the remaining nodes would act as independent pacemakers. But they would run the heart too slowly, and the patient would need a pacemaker. (SA and AV nodes can act as pacemakers.) That is, if the patient suffers symptomatic bradycardia, they would need an artificial pacemaker.

The currents that move through the heart are ionic currents, waves of depolarization and repolarization that move through the neural tissue and cardiac muscle (different from smooth or striated muscles). The wave that moves through the neural tissue is a wave of sodium ion depolarization followed by potassium ion repolarization.

Sympathetic Nervous System

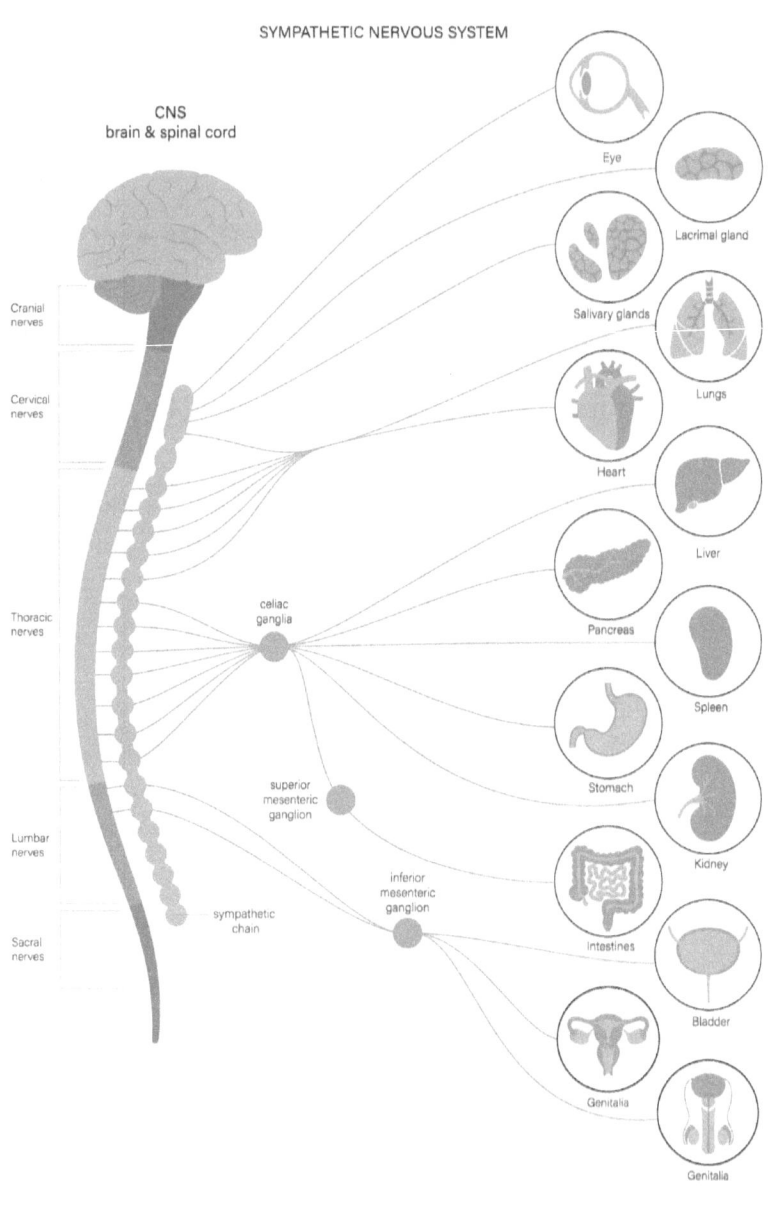

Currents that disseminate across cardiac muscle do so by calcium depolarization:

In a healthy heart, neural impulses will start at the SA node, progress down toward the AV node, and then progress toward the apex at which point they will change direction as they follow the Purkinje fibers. So if we record them on a piece of paper we will see:

Q: movement of ionic current from AV node to apex, through the bundle of His (intraventricular wall)

R: ionic current reverses direction as it travels through the Purkinje fibers (in the lateral walls of the ventricles)

S: heart repolarizes after ionic current has finished its course.

EKGs are divided into two parts:

The limb leads aVR, aVL, AvF, I, II, and III

The chest leads are V1, V2, V3, V4, V5, and V6

When we read EKGs, there are twelve questions that we want to answer:

Does the heart have a regular rate and rhythm?

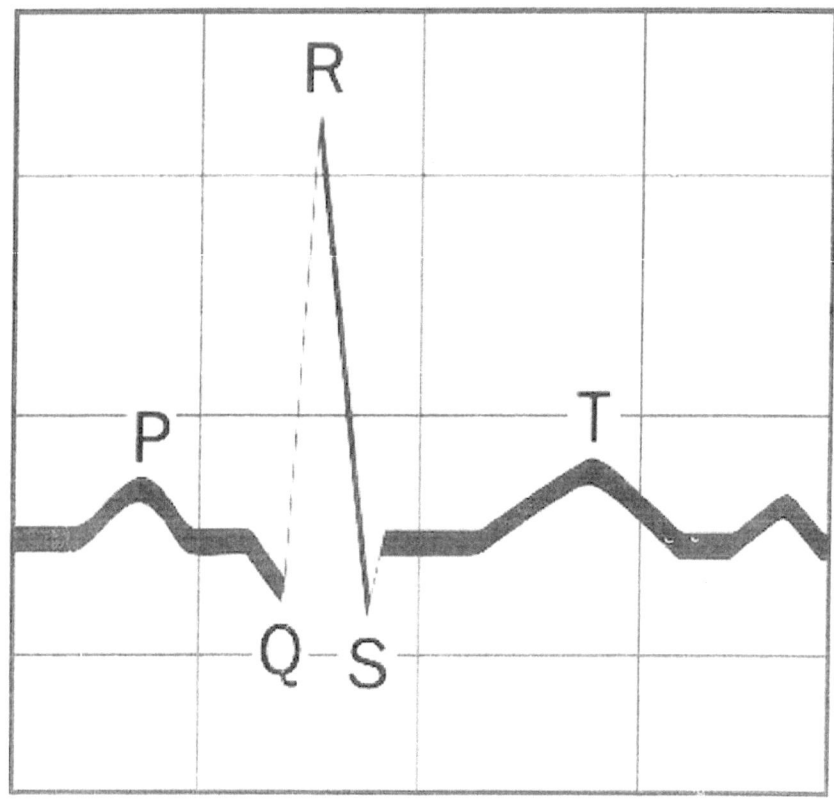

QRS Complex

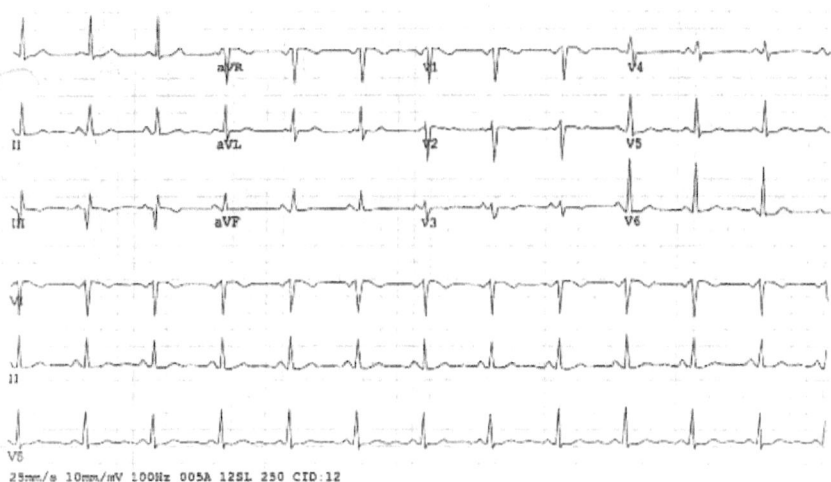

25mm/s 10mm/mV 100Hz 005A 12SL 250 CID:12

EKG

Did the patient have a heart attack (MI) in the distant past (over three months ago)? The patient might not know.

Does the patient have atrial flutter or fibrillation?

Do they present with sinus tachycardia (rapid heart rate)?

Has the AV node been damaged?

Has the bundle of His been damaged?

Is the heart receiving enough oxygen?

Has there been recent damage to the heart from anoxia (lack of oxygen)?

Is the blood pressure in either ventricle too high?

Does the patient have Wolff-Parkinson-White syndrome?

Is medication affecting the ability of the heart to repolarize?

Is the heart putting out a low voltage? This could be caused by:

- Hypothyroidism

    Emphysema

    Pericarditis

When reading an EKG the interpreter will begin by taking a pair of calipers and "marching" the QRS complexes out. The distance between each should be equal, and there should be one QRS complex per second. If they do not march out, you would say that the heart had an irregular rhythm. We will discuss the possible causes later.

To know if someone had an MI over three months ago, we would look at the limb leads: aVR, aVL, and aVF. If we see Q waves—for example:

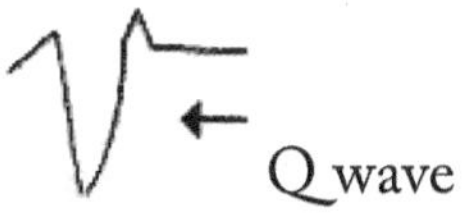

—this could alert us to the fact that they might have had an MI in the distant past. This is something that patients are often unaware of. They might have had the attack during their sleep or they might have mistaken it for indigestion. A suggestion of a heart attack should prompt a full cardiac workup by a specialist. They should see a cardiologist at least once.

Frederick Wenckeback 1906

Another possibility is that they could be diabetic. Diabetics can have an MI without "feeling" it. It will not cause pain or discomfort.

If there are multiple P waves between each QRS complex, the patient might have atrial flutter. If the flutter is ongoing and so fast that you cannot see P waves, the patient might be suffering from atrial fibrillation. Both conditions could be caused by damage to the atria and/or by oxygen deprivation. Either could cause a blood clot and could be deadly.

It is occasionally possible to stop atrial fibrillation or flutter. But most patients are placed on medications such as Eliquis (which is safer than Coumadin, an older anticoagulant). These medications could lower the chances throughing a clot to the brain five times.

If the investigators observe normal P waves and QRS complexes, but the heart is running either too fast or too slow, it would be indicative of sinus tachycardia or sinus bradycardia. It would represent too much influence from either the sympathetic nervous system or the parasympathetic. Tachycardia could be treated with metoprolol extended-release or clonidine. Bradycardia is usually treated with a pacemaker.

Wenckebach

Suppose you look at an EKG and see one of three conditions:

- The interval between the P waves and the QRS is too long.
- Or the interval gets longer with each cycle, then a P wave is dropped.
- Or there is complete dissociation between P waves and QRS complexes.

This would be:

- First-degree Wenckebach (which does not require treatment)
- Second-degree Wenckebach (which requires a pacemaker)
- Third-degree Wenckebach respectively (which requires a pacemaker)

Suppose we look at an EKG and see notching of the QRS? This would indicate that the bundle of His had been damaged: probably by an MI.

Double notching:

If we look at an EKG and see S-T elevation by more than 2 mm, it would suggest ischemia: lack of oxygen.

S-T elevation by more than 2 mm.

If someone has ischemia, they should take one aspirin 81 mg by mouth and place a sublingual nitroglycerine under the tongue every five minutes three times or until the pain stops. They should be hospitalized, see a cardiologist, and be started on a long-acting nitrate such as isosorbide mononitrate.

If we see 2 mm of S-T depression, it would suggest that damage has already started.

To prevent further damage, the patient should:

- Lower their blood pressure (to less than 140/90).
- Take medication to reduce cholesterol.

  - ➤ HDL should be above 40 mg/dL.
  - ➤ LDL and VLDL should be below 100 mg/dL.

This usually requires a statin such as Lipitor or Crestor.

Control diabetes (if your hemoglobin A1C or HgA1C is less than 10, you could probably control it with oral medication such as metformin, Januvia, Glucovance, Glyburide, Onglyza, or Actos). If it is above 10 you might need insulin.

Take aspirin 81 mg a day.

Exercise four times a week. Do aerobic exercises for at least twenty minutes a session. This includes jogging, biking, swimming, hiking, tennis, or golf. They need to increase their heart rate by 20%.

Stop smoking. Your caretaker could help by offering Chantix, nicotine patches, or nicotine gum. Vape cigarettes are questionable.

Slow the heart and decrease its demand for oxygen with beta blockers such as Metoprolol, Atenolol, or Labetalol.

Suppose we look at an EKG and see notching of the QRS complex.

This represents damage to the bundle of His, which could be caused by an MI. You would use the same precautions to prevent a recurrence.

If the conduction system were severely damaged, you might see a wide QRS complex:

Suppose we see "low voltage" (small amplitude on an EKG). This could be:

- Hypothyroidism

  Emphysema

  Pericarditis

And the QRS complexes in V1, V2, and V3 should be down facing.

Enlargement of V1, V2, and V3 would suggest cor pulmonale (enlargement of the right ventricle). Cor pulmonale is usually caused by pulmonary hypertension: anything that causes blood pressure to go up in the lungs.

If the QRS complexes in V4, V5, and V6 are large, it would suggest cardiomegaly (enlargement of the left ventricle).

Cardiomegaly could be caused by:

Systemic hypertension (which would be diagnosed on a physical exam)

Aortic stenoses (limited mobility of the aortic valve; would appear on an echocardiogram)

IHSS (hypertrophy of part of the left ventricle; would also appear on echocardiogram)

In both cases the heart would have to work too hard and would build up excess muscle. This would cause it to generate too much electricity and would lead to an enlarged QRS complex.

In biology, electricity comes from muscle. The electric eel (which is not a true eel) has a modified muscle that runs through its body.

Another way to look at V1–V6 is in terms of axis. Its direction should change between V3–V4.

If the neural pathways were damaged, the impulses might be forced to flow through cardiac muscle. They would do so slowly, and it would present as a wide QRS complex. Such patients should receive MI prevention and nitrates: initially sublingual nitroglycerine. Then they should be placed on long-acting nitrates such as isosorbide mononitrate. An angiogram should be performed by a cardiologist, who might place stents or recommend bypass surgery.

If one reads an EKG and observes the opposite, S-T depression (of more than 2 mm), it would suggest that damage had already started and that intervention was critical.

There are two additional topics that we need to discuss. As we progress from V1 to V6, we would see two things:

> The current would reverse direction at V3 and V4. This is referred to as the "axis." As we move from V1 to V3 we will see the right ventricle and the current traveling from the AV node to the apex. As we move from V4 to V6, we see the left side of the heart (as the ionic current approaches the ends of the Purkinje fibers).

Another condition, seen in 1 to 3 individuals out of 1,000, is Wolff-Parkinson-White (WPW), which can kill an athlete. It was worked out by three physicians in 1930 and presents as a U-shaped wave before each QRS. It represents reentry current. Current travels through the bundle of His and the Purkinje fibers. Then it passes back to the AV node through aberrant conduction tissue. If such a patient were to exert themselves, they could go into cardiac arrest.

It is a rare condition. But if it is present in a young athlete, it can kill them.

When examining athletes, it is important to ask:

- Have you played in the past? If so, did you have any medical issues?

 Did anyone in the family die suddenly before the age of 40?

 Did any family member experience an MI before the age of 55?

If the answer to any of the above is yes, they should have an EKG. WPW would present as a U-shaped wave before the QRS complex, and they should be seen by a cardiologist. It can be corrected with medication or with electrical ablation.

Suppose the investigator sees small QRS complexes. This could be caused by either:

- Pericarditis in association with SLE; this could be worked up with an EKG, a blood lab (for SLE), and an echocardiogram; or

- COPD with air being trapped in the pericardial sac and acting as an insulator, which could be worked up with a chest X-ray PA and lateral and a pulmonary function test; or
- Hypothyroidism, which could be worked up with a blood lab (TFT).

Suppose the investigator sees an S-T segment that is too long. This would suggest that something is slowing down the repolarization of the heart such as erythromycin interacting with something else. It could be evaluated by running the patient's medications through a pharmaceutical program.

Wolff-Parkinson-White 1930

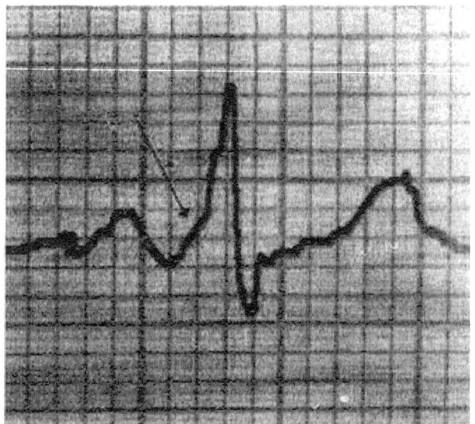

U-shaped wave precedes the QRS complex.

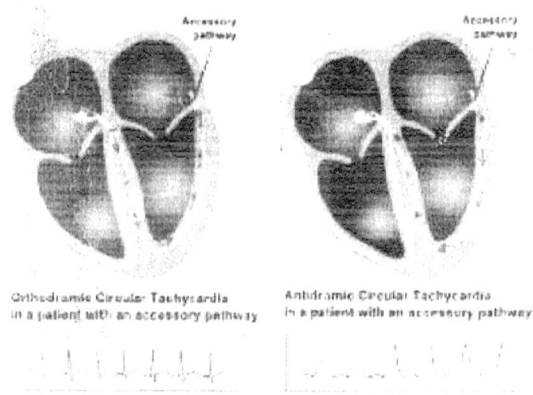

Reentry current in 1–3 patients per 1,000

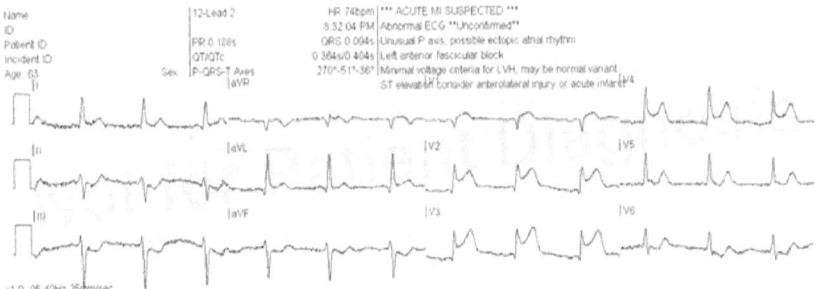

Wolff-Parkinson-White

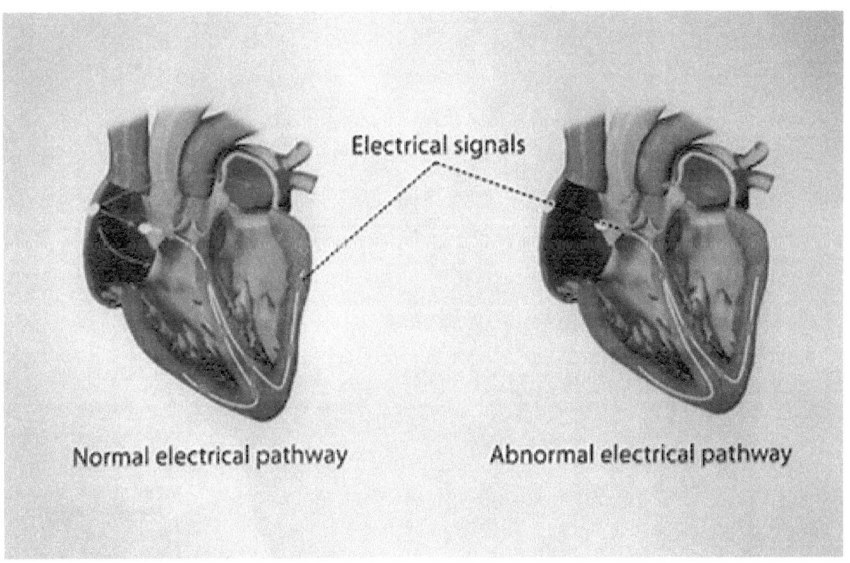

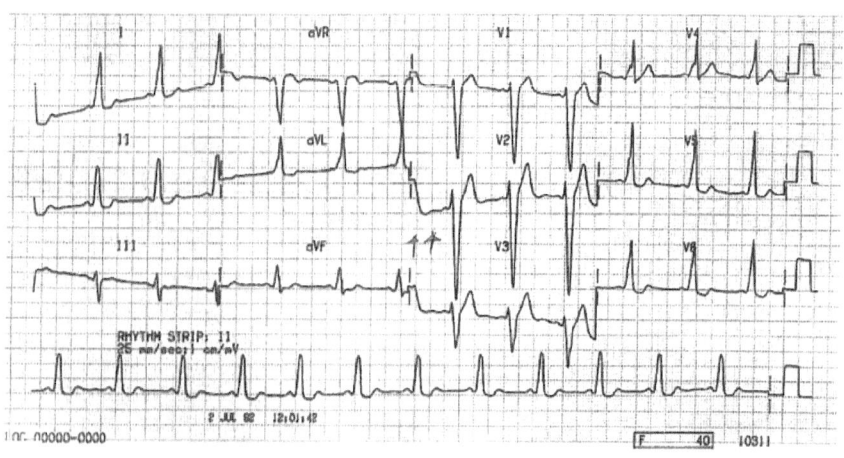

Heart

EKG

Let us talk a little more about EKGs. You might be wondering what the difference is between the limb leads (I, II, III, aVR, aVL, and aVF) and the chest leads (V1, V2, V3, V4, V5, and V6)? The difference is that they allow us to rotate around the heart in two planes.

With the limb leads we rotate around the heart in a coronal plane (a plane that divides the body into anterior and posterior). We start at a theoretical point above the head and go around the body until we end up where we started: aVR and aVL refer to the left and right limbs. I, II, III, and aVF are theoretical points between the limbs. We can use limb leads to determine the rate and rhythm, the PR interval, and to see if the patient had an MI more than three months ago.

The chest leads allow us to rotate around the heart in a sagittal plane, dividing the body into left and right segments. We start on the right-hand side of the chest (V1) and turn counterclockwise toward the left (V6). We use this to answer all the other questions that we wish to address.

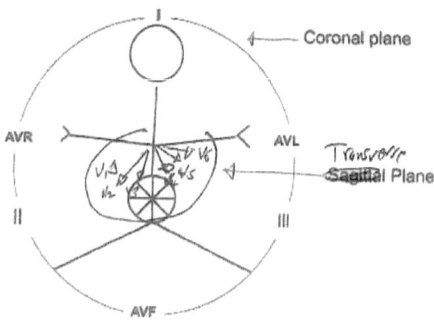

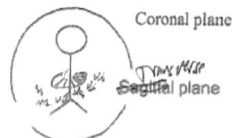

Go around the body in a coronal plane.

Go around in a sagittal plane.

Rita Levi-Montalcini

Lungs

Let us move onto another organ the lungs. Lungs arise from mesoderm, but the lining of the air passages comes from ectoderm. Bronchi and bronchioles are conduits that allow air to enter the alveoli. They consist of walls that are 1/10,000 of an inch thick and allow oxygen and carbon dioxide to diffuse into and out of capillaries that circulate blood.

During early development lung buds grow out of the mediastinum in the area of the main bronchi. When they touch ribs and the diaphragm, electrochemical messengers cause "contact inhibition," and two things happen:

- The lungs stop growing.

    They differentiate into their adult form.

A former member of the Italian senate, Rita Levi-Montalcini, won a Nobel Prize for her work on electrochemical messengers.

The respiratory system consists of the:

1. Trachea
2. Carina
3. Bronchi
4. Bronchioles
5. Alveoli

On the left there is an inferior and a superior bronchus.

On the right there is an inferior, a middle, and a superior bronchus.

The walls of the alveoli are about a 1/10,000 of an inch thick and are surrounded by capillaries that carry deoxygenated blood. Oxygen diffuses through the wall into the capillary and is taken up by red blood cells. These cells also release carbon dioxide that diffuses back into the alveoli and is exhaled.

The lungs expand when the chest cavity expands, and vice versa. They collapse when the chest cavity collapses. The chest cavity is driven by three groups of muscles and their associated nerves.

> The phrenic nerves run along the cervical and thoracic spines and innervate the diaphragm. When they contract, they create space within the chest cavity. The nerves arise from c3, c4, and c5. Remember the mnemonic: "3, 4, and 5 keeps the diaphragm alive."

> The sternocleidomastoids also create space by pulling the ribs up. They are innervated by the spinal accessory nerve (XI) that originates in the XI cranial nucleus one on either side of the brain.

> The intercostal muscles between the ribs help to contract the chest cavity. They are innervated by the nerves that course with the ribs, VAN (vein, artery, nerve).

The lungs are driven by a part of the brain known as the medulla oblongata ("oblong shaped middle object" in Latin), which is responsible for most life functions (such as pumping of blood). If it senses that blood oxygen is low, it will speed the lungs up. If it senses that blood carbon dioxide is high, it will do the same. If it senses that blood pH is too low, it allows the lungs to blow off carbon dioxide. If it senses that serum pH is high, it will slow them down so as to retain carbon dioxide.

Let us conclude by considering some of the disease processes that affect the lungs.

Pneumonia can be caused by:

- Bacteria

  Viruses
  Fungi
  Parasites
  Chemical irritants

Pulmonary Function Tests:

Pulmonary function tests can be used to work up pulmonary conditions. Does the patient have normal lungs? Asthma? Or emphysema or pneumonia? You want to know early.

ml. of inhaled air:

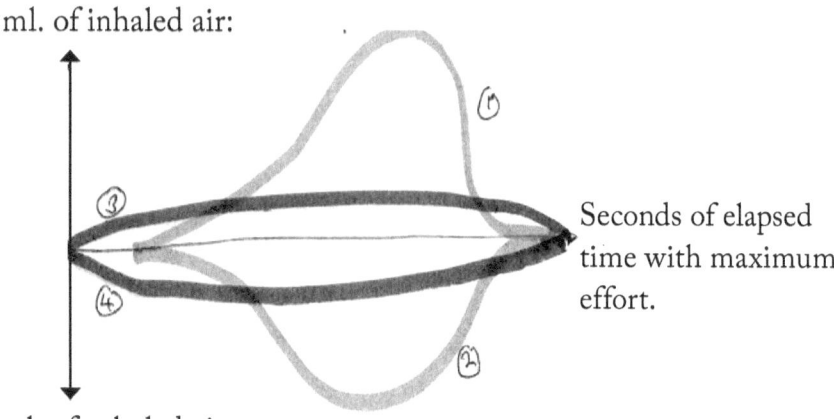

Seconds of elapsed time with maximum effort.

ml. of exhaled air:

[1&2] Normal respiratory effort. The patient can inhale and exhale a large volume of air in a short period of time.

[3&4] If the patient suffers from emphysema or asthma, it will take them a long time to inhale and exhale a small volume of air.

If we discover that the patient has a restrictive pulmonary disease (i.e., the blue curve), the next step would be to know if it is emphysema or asthma. So we would give them a bronchodilator (such as albuterol) and test them again. Asthma would resolve, and their pulmonary function curves would look normal. If it remained compromised, you would think of emphysema. But supposing they did not have a history suggestive of emphysema (i.e. they did not smoke), you would do two things:

1. Obtain a chest X-ray PA/Lat
2. Obtain a pulmonary biopsy

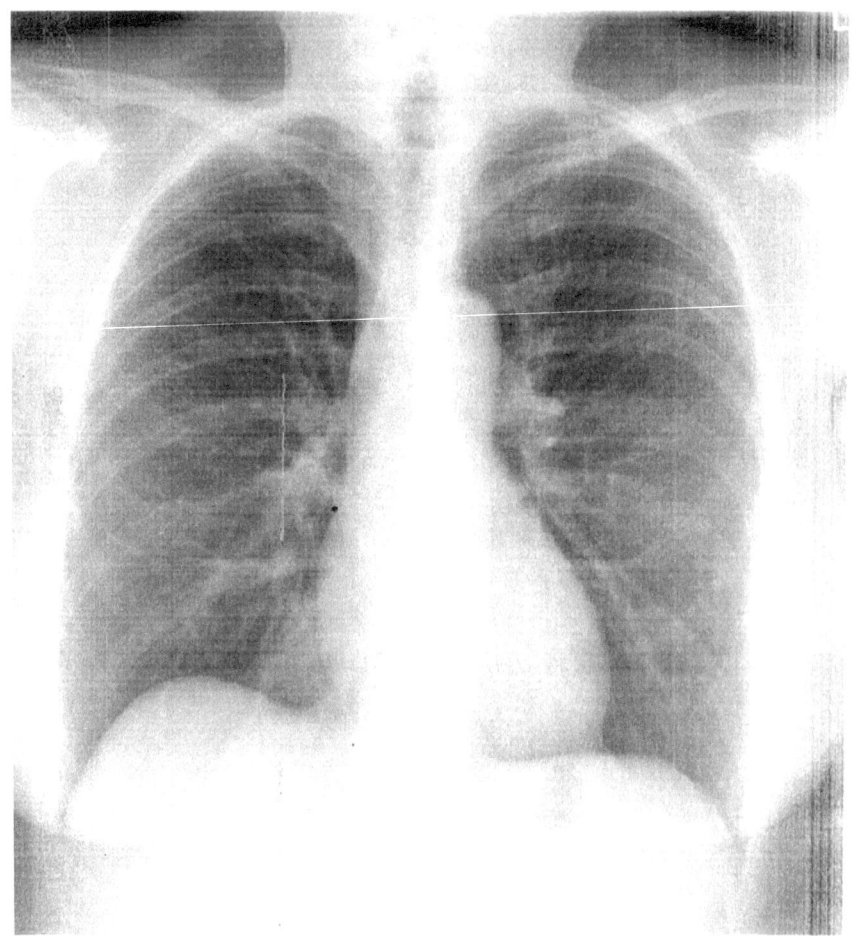

Normal chest X-ray.

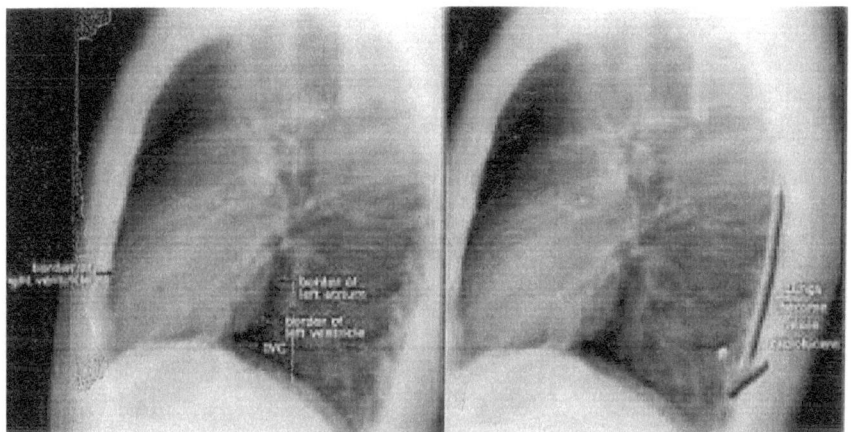

Normal chest X-ray.

Table 3

Severity of airflow obstruction based on percentage (%) predicted forced expiratory volume in 1 second (FEV1).

FEV1 % predicted	Stage
>80%	Mild
50–79%	Moderate
30–49%	Severe
<30%	Very severe

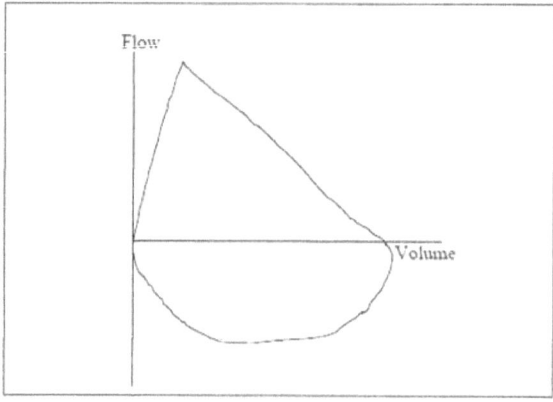

Fig. 5

Flow volume curve in obstructive lung disease, e.g., COPD

With knowledge of the expected appearance of the flow volume loop in a normal patient, important information can be obtained from the morphology of the curve in patients with suspected respiratory disease. Patients with obstructive lung diseases with reduced expiratory flow in the peripheral airways typically have a concave appearance to the descending portion of the expiratory limb (Figure 5) rather than a straight line. In patients with emphysema the loss of elastic recoil and radial support results in pressure dependent collapse of the distal airways with more pronounced "scalloping" of the expiratory limb. Even if the flow volume loop morphology is normal, a reduction in PEFR may be an indication of asthma with early airways obstruction. Similarly a reduction in FEF25-75% indicates small airways obstruction. This can also occur in patients with asthma with a normal PEFR, and is useful in providing a better overall picture of asthma control. It is also helpful in monitoring response to treatment and this may be particularly important for patients being considered for general anaesthetic and surgical intervention.In restrictive defects the expiratory limb has a convex or linear appearance because flow rates are preserved but the problem relates to a parenchymal disorder e.g. lung fibrosis which reduces lung volumes.

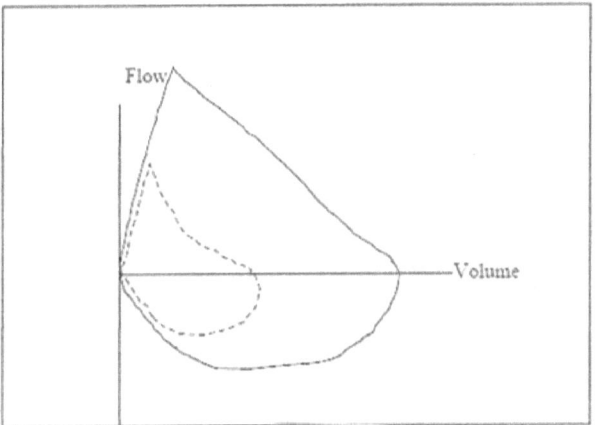

Fig 5

Flow volume curve in obstructive lung disease e.g. COPD

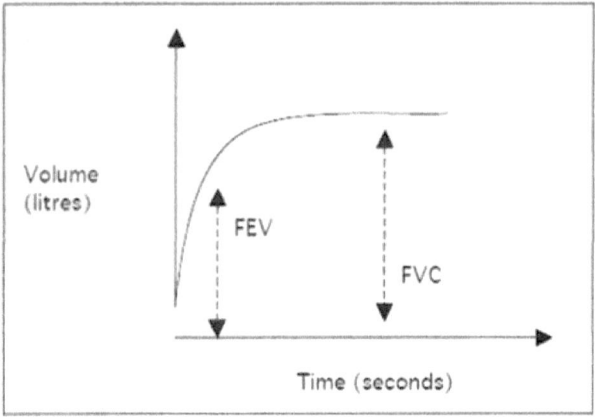

Fig 1

Normal Spirometry

- Forced expiratory volume in one second (FEV1)
- Forced vital capacity (FVC)
- The ratio of the two volumes (FEV1/FVC)

Spirometry and the calculation of FEV1/FVC allows the identification of obstructive or restrictive ventilatory defects. A FEV1/FVC < 70 % where FEV1 is reduced more than FVC signifies an obstructive defect (Figure 2). Common examples of obstructive defects include chronic obstructive pulmonary disease (COPD) and asthma. The FEV1 can be expressed as a percentage of the predictive value which allows classification of the severity of the impairment (table 3)[8]. An FEV1/FVC > 70% where FVC is reduced more so than FEV1 is seen in restrictive defects such as interstitial lung diseases (e.g. idiopathic pulmonary fibrosis) and chest wall deformities (Figure 3).

If it is pneumonia, it can be subsequently treated with antibiotics, antivirals, antifungals, or anthelminthic, or by withholding the chemical irritant.

Emphysema is a loss of elasticity by the alveolar sacs. In more advanced states the bronchioles also lose elasticity. In its early presentation it will be difficult for the patient to get air out of the alveolar sacs. They acquire a pink complexion and work hard to breathe. They are sometimes referred to as "pink puffers."

As the condition becomes more advanced, bronchioles also begin to collapse, and air is trapped in the alveoli. These patients develop a cyanotic appearance and will appear to have inflated lungs. They are sometimes referred to as "blue bloaters."

COPD is chronic obstructive pulmonary disease and is an advanced stage of emphysema.

In its early stage, emphysema can be treated with:

- Bronchodilators and inhaled steroids.

  Humidified air

  Oxygen

In its more advanced state, the patient might require portable oxygen and/or a heart/lung transplant.

With pulmonary fibrosis, it becomes difficult for carbon dioxide and oxygen to pass through the alveolar wall. This can be caused by certain medications such as methotrexate, but other causes are largely unknown. It would be diagnosed with a pulmonary biopsy.

In its early stage, it can be treated with bronchodilators, inhaled steroids, and oxygen. In its more advanced stages a lung transplant may be necessary.

Asthma is a condition in which bronchial walls spasm in response to:

- Exercise

    Cold air
    Chemical irritants
    Allergens
    Emotional stress

It is treated with:

- Bronchodilators and inhaled steroids

    Humidified air

    Air with a high concentration of oxygen

The Digestive System

Let us move on to the digestive system. This can be divided into ten parts:

- Mouth

  Oropharynx
  Esophagus
  Stomach
  Small intestine
  Gallbladder
  Liver
  Pancreas
  Large intestine
  Rectum

The mouth contains 32 teeth—eight molars, eight premolars, eight canines and eight incisors. The incisors cut the food while the canines hold it. The premolars grind soft food, and the molars grind hard food.

The mouth also contains the tongue, which is capable of distinguishing five basic tastes—salty, sour, sweet, bitter, and rancid—and is capable of distinguishing 7 million different taste grades. The taste buds are on the sides and the top of the tongue.

The tongue consists of the glossopharyngeal muscle and is innervated by the ninth cranial nerve (responsible for taste) and the twelfth cranial nerve, which extends it. The seventh and tenth cranial nerves also help in the appreciation of taste.

The mouth is supplied with saliva by the

- Parotid glands that are within the cheeks, and

  Sublingual glands that are in the floor of the mouth.

The parotid glands secrete saliva through Stenson's ducts. They are superior to the molars on the hard palate, one on the left and one on the right.

The sublingual glands secrete fluid through a duct that is below the tongue (embedded in the frenulum).

The principal blood supply to the tongue is a circular-shaped vessel known as the lingual artery, which branches off the external carotid.

The human mouth contains a wide variety of flora and is said to be "more dirty" than a dog's mouth. But saliva acts as an antiseptic and contains the enzyme amylase, which breaks starch down to glucose. And so digestion begins from the moment food enters the mouth.

We then use the muscles of the tongue and the oropharynx to guide food into the esophagus. While we swallow, it is necessary to close the epiglottis. This prevents food and liquids from entering the lungs.

The esophagus is a long muscular tube that extends from the oropharynx to the upper part of the stomach just below the diaphragm. The inner surface of the esophagus is lined with epithelium. It receives blood from the carotids and the aorta, and its neurological innervation comes in part from the Auerbach plexus, a kind of nerve net that surrounds the esophagus and the rest of the GI tract. It creates a successive volley of impulses that leads to a wave of muscular contraction known as peristalsis. So when you distend the upper part of the esophagus it creates a wave of muscular contraction that travels the length of the GI tract.

The esophagus is a flexible structure whose contractions propel food and liquid from the oropharynx to the stomach. Just above the stomach there is a sphincter, which helps keep liquid contents within the stomach. It contracts so that we will not be vomite (which works most of the time).

Once inside the stomach, contents are mixed with hydrochloric acid and an enzyme called pepsin. Pepsin is a protease that is activated by low pH and assists in the digestion of meat. It is synthesized by Chief cells in the wall of the stomach. At low pH, pepsinogen becomes pepsin, which is its active form. HCl (hydrochloric acid) is synthesized by cells in the wall of the stomach. HCl also helps in the absorption of the B vitamins (which are absorbed by intrinsic factor, which is also activated by low pH) and minerals, such as iron, calcium, magnesium, and zinc. We need HCl to kill dangerous bacteria and parasites.

Known risk factors for stomach cancer include foods that are rancid (and contain nitrites), burned, or smoked; smoking; non-healing ulcers (3 percent chance), and achlorhydria (lack of HCl acid). Some research also suggests that extremely salty foods could be a risk.

The stomach receives blood from two sources:

The gastric artery which branches off the gastrosplenic artery and runs along the stomach's greater curvature, and branches off the intestinal artery that travel with the intestines. The splenich artery also comes off of the gatrospenic. Its neurological innervation comes from the Auerbach plexus and nerve roots that travel with the mesentery (a suspensory ligament that anchors it to the spine).

When food leaves the stomach it enters the small intestine. The small intestine can be divided into three parts:

- Duodenum

  Jejunum
  Ilium

The pH of the small intestine is greater than seven. It receives a liquid called bile which is green and contains sodium bicarbonate. It enters the duodenum through the Ampulla of Vater, an orifice on the wall of the duodenum which is connected to the hepatocystic duct. The hepatocystic duct is formed by the union of the hepatic duct that comes from the liver and the cystic duct that originates from the gallbladder. When we eat, the gallbladder contracts and pumps bile into the duodenum. This helps us to digest fats. Particles of fat are first emulsified (turned into very small particles). They are then acted upon by sodium bicarbonate and lipase, which break them down into their basic components, glycerin and long-chain fatty acids. The bile is synthesized in the liver and stored in the gallbladder. The surface of the intestine also contains villi, meanming "fingers" in Latin. Villi increase the absorptive surface area.

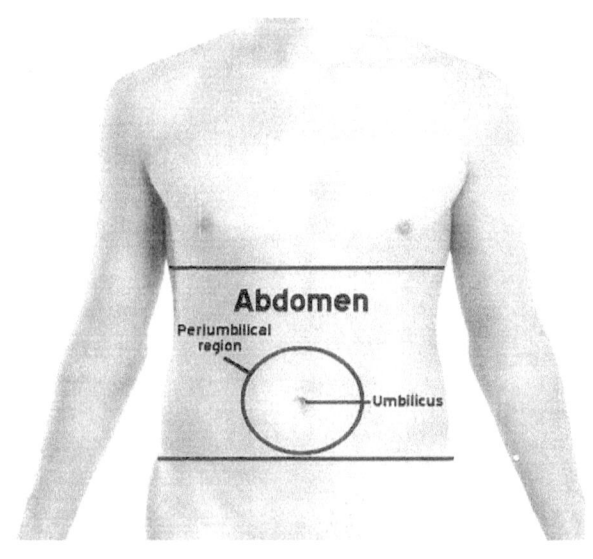

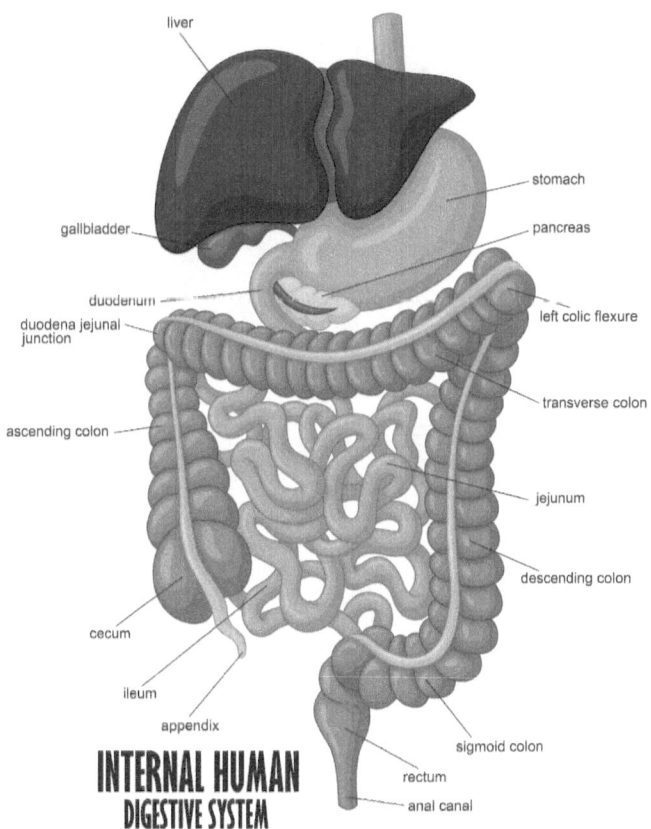

INTERNAL HUMAN
DIGESTIVE SYSTEM

Intestine Diagram

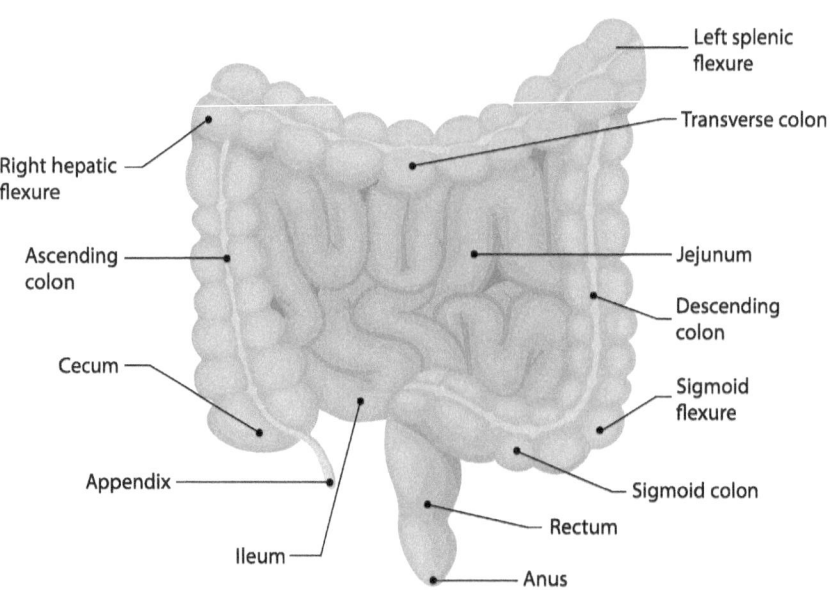

Left splenic flexure

Transverse colon

Right hepatic flexure

Ascending colon

Jejunum

Descending colon

Cecum

Sigmoid flexure

Appendix

Sigmoid colon

Rectum

Ileum

Anus

Small intestine components

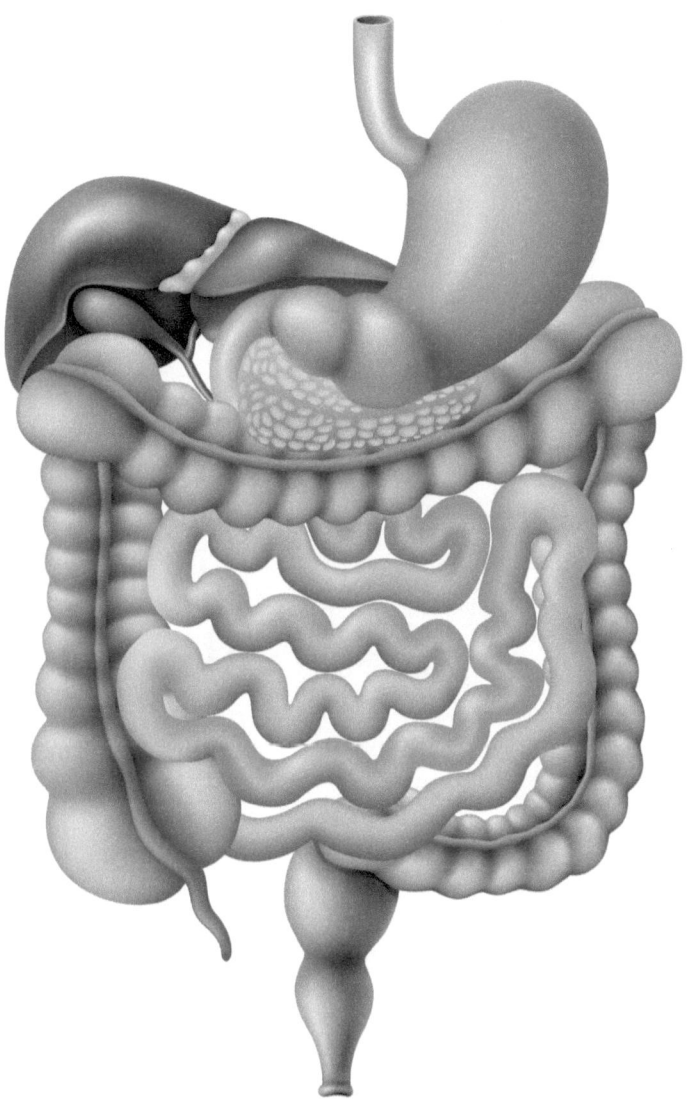

Gastrointestinal Organs

Digestive System

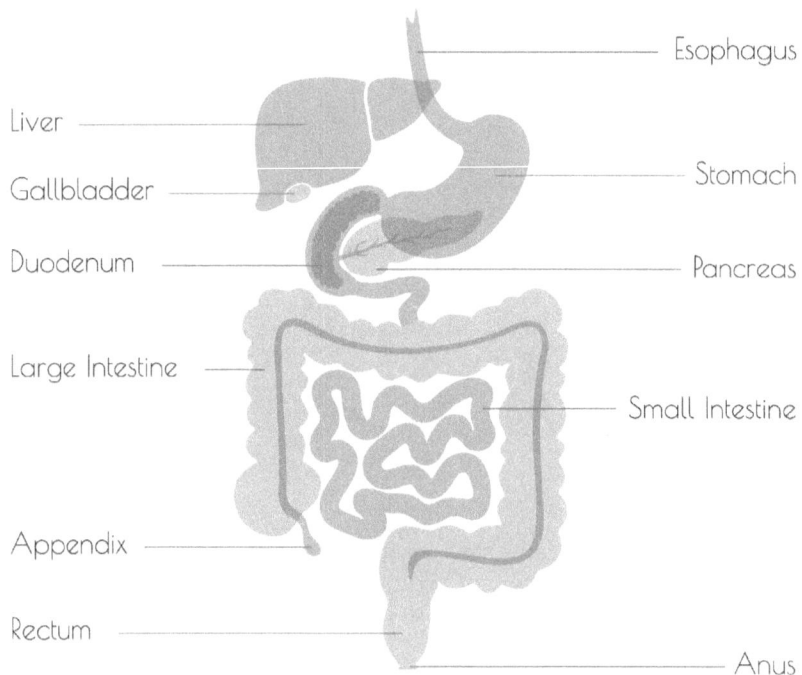

Esophagus

Liver

Gallbladder

Duodenum

Large Intestine

Appendix

Rectum

Stomach

Pancreas

Small Intestine

Anus

Abdominal organs

Let us talk briefly about the risk factors for stomach cancer. We will discuss the staging and grading of tumors later. As a general rule, stomach cancer can be avoided by:

- Not smoking

  Not eating burned food
  Not eating spoiled food
  Not eating smoked foods
  Allowing all ulcers to heal. This usually happens in six weeks with good care.

The best way to avoid colon cancer is to:

- Not smoke

  Avoid red meats
  Eat vegetables and fruits
  Eat green leafy vegetables such as lettuce
  Maintain normal weight
  Control diabetes and exercise

Start colonoscopies at the age of 50. (Cologuard is acceptable, though not as accurate.)

The absorptive surface area of the large intestine (colon) is increased by ridges and grooves in the wall. Their main function is to absorb water so that we do not suffer perpetual diarrhea or dehydration.

There is one final topic that deserves discussion: hemorrhoids. Hemorrhoids are veins that have partially (or totally) occluded. Once occluded, they become engorged with blood. They may or may not become irritated and/or bleed. There are two kinds of hemorrhoids: external and internal. To understand this we have to define another term: dentate line.

If an investigator performs a colonoscopy they would see the dentate line soon after entering the rectum. It would appear approximately 6 cm distal to the anal verge. The dentate line can be characterized in three ways.

It receives blood from the rectal arteries which branch off the iliac. It also receives some blood from the intestinal artery.

It is innervated by the splanchnic nerves, which arise from the sacral vertebrae. These are mostly parasympathetic nerve fibers and the lymphatics drain into the lymph nodes that accompany the intestines and rectal veins.

It does not feel pain.

Staging, Treatment, and Five-Year Survival with Colon Cancer

Stage:	Treatment:	Five Year Survival:
Stage 0 Carcinoma in situ confined to inner lining of colon	Surgical resection of tumor. Sample at least 12 regional nodes.	91%
Stage I Tumor invades submucosa and/or muscularis propria	" "	"
Stage IIA Tumor penetrates subserosa of colon.	Consider downstaging tumor with preoperative chemo and/or radiotherapy prior to resection	"
Stage IIB Tumor penetrates the peritoneum that surrounds the colon	Consider postoperative radiotherapy for surgical margins of less than 2 cm and/or for poorly differentiated tumors.	"
Stage IIC Tumor spreads to adjacent organs and/ or to regional nodes (1–7)	Gross resection of all visible tumor, lymph node sampling, and post-operative chemotherapy. At least 12 regional nodes should be biopsied.	72%

Stage:	Treatment:	Five Year Survival:
Stage IIIA tumor has spread to local organs and nodes.	Downstage bulk tumor with preop. radiotherapy. Resect bulk disease and add postoperative chemotherapy.	72%
Stage IV Distant metastases.	Resect all gross disease locally and metastases. Use cryotherapy for unresectable lesions. Preoperative radiotherapy to downstage bulky lesions. Postoperative chemotherapy. Postoperative radiotherapy for margins closer than 2 cm, poorly differentiated tumors, or for the palliation of pain.	14%

Let us take a minute to consider "the first pass effect." All substances that leave the GI tract go to the liver first. This is the "first pass" effect. The liver will detoxify some of them such as ethyl alcohol.

But other substances such as lipids can accumulate in hepatocytes (cells of the liver). Lipids are consumed by hepatocytes through phagocytosis. Any tumor that develops in the GI tract will spread to the liver when it is advanced.

The rectum proximal to the dentate line (near the anus) is lined with squamous epithelium. The rectum distal to the dentate line is mostly glandular (adenosquamous).

External hemorrhoids are innervated by the nerves that accompany the skin dermatomes. It also receives its blood supply and its lymphatic drainage from vessels that accompany the dermatomes.

If a hemorrhoid hurts or itches, it's external. If it does not, it is internal. Tumors will metastasize with the intestinal nodes or with structures in the skin depending upon whether or not they occur above or below the dentate line. So if a tumor starts above the dentate, it would metastasize to pelvic nodes. If it starts below it would spread with the skin.

The reader is referred to a standard text for ways to treat hemorrhoids, although it is fairly simple.

The Central Nervous System

Let us move on to a different system: the central nervous system, which consists of:

- Brain

  Spinal cord
  Parasympathetic and sympathetic nervous systems
  Peripheral nerves

The brain forms from the neural tube. When higher organisms develop, they start with a flat plate of epithelium which runs the length of their body. As the two ends move away from each other the edges of this plate curl upward and form a neural groove which eventually closes to form a neural tube. The cephalic (which means "head" in Latin) end of this tube begins to bulge outward and differentiates into the brain. Parts of this new brain called the telencephalon extend forward like the eyes of a snail, until they touch the skin of the face. At that point the skin differentiates into an eyelid, cornea, and lens. The stalk differentiates into a retina, a macula, a fovea, an optic nerve, the chiasm, and the optic tracts. The rest of the brain differentiates into many structures to include:

Frontal lobes: responsible for emotion

- Broca's area: responsible for speech

Olfactory lobes: responsible for sense of smell

Lateral lobes: responsible for hearing

Uncinate: responsible for short-term memory

Cranial nuclei: responsible for the activity of the 12 cranial nerves

Substantia nigra: responsible for voluntary movement (the patient will develop Parkinson's disease if this area is damaged). "Substantia nigra" means "the black substance" in Latin. It looks like a black anchor at the center of the brain.

Parietal lobes: responsible for long-term memory, emotionality and thought

Precentral gyrus: responsible for voluntary muscular activity

Postcentral gyrus: responsible for sensation

Chordate nuclei (part of a feedback system for the cortex)

Pituitary gland: responsible for most hormonal activity

The pituitary plays a key role in menstrual periods. During the first half of the cycle the anterior lobe of the pituitary secretes follicle-stimulating hormone (FSH), which stimulates ovulation and estrogen formation by the ovaries. Estrogen causes the uterus to create endometrial lining with glands and blood vessels in preparation for a zygote. During the second half the anterior pituitary secretes luteinizing hormone (LH) if there is no zygote. This causes the ovaries to develop corpus luteums that are yellow (because of lipids) and secrete progesterone. Progesterone causes exfoliation ("shedding of leaves" in Latin), in which the lining breaks down and initiates the menstrual period.

Hypothalamus: influences the pituitary

Thalamus: hunger, emotionality, and many other influences

Pineal gland: additional hormonal activity and the secretion of melatonin. Also carries out the circadian rhythms. They are a "natural clock."

Occipital lobes: responsible for vision

Medulla oblongata: drives the body's core functions, pumping blood, breathing, and other vital processes

Brainstem: connects brain to the spine

Cerebellum: responsible for muscular coordination

Red nuclei: vital part of muscular activity

Brain Anatomy

Adapted from *Human Anatomy and Physiology*
by Marieb and Hoehn (9th ed.)

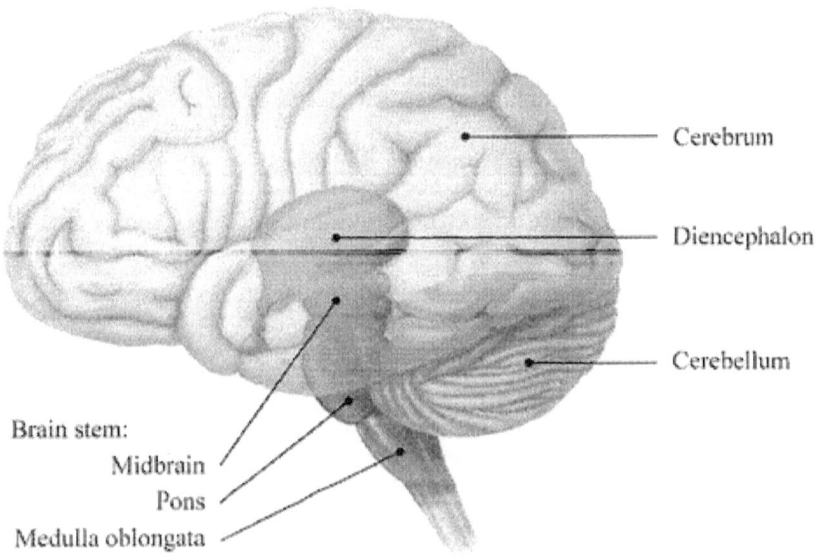

Figure 1: General anatomy of the human brain
Marieb and Hoehn
(*Human Anatomy and Physiology*, 9th ed.) – Figure 12.2

CEREBRUM

Divided into two hemispheres, the cerebrum is the largest region of the human brain—

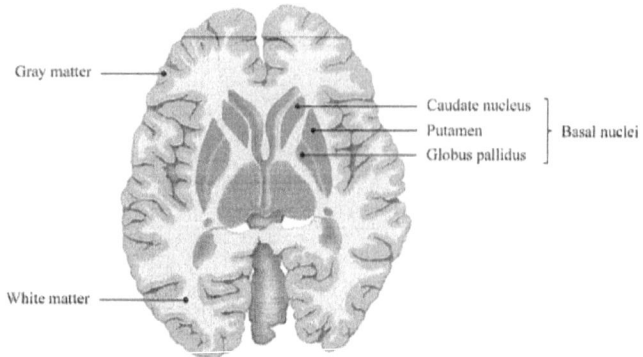

Figure 2: Transverse section of cerebrum showing
major regions of cerebral hemispheres
Marieb and Hoehn
(*Human Anatomy and Physiology*, 9th ed.) – Figure 12.9

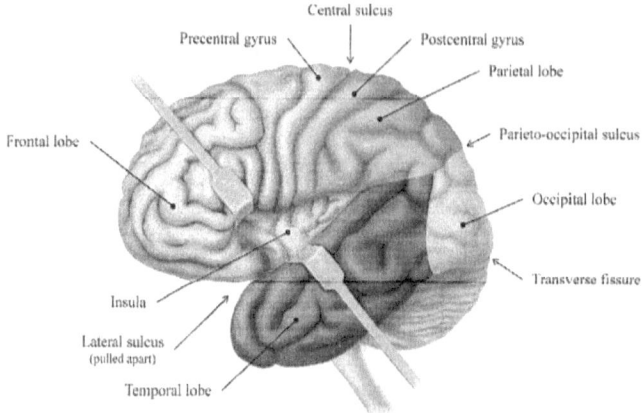

Figure 3: Lobes, sulci, and fissures of the cerebral
hemispheres (longitudinal fissure not pictured)
Marieb and Hoehn
(*Human Anatomy and Physiology*, 9th ed.) – Figure 12.4

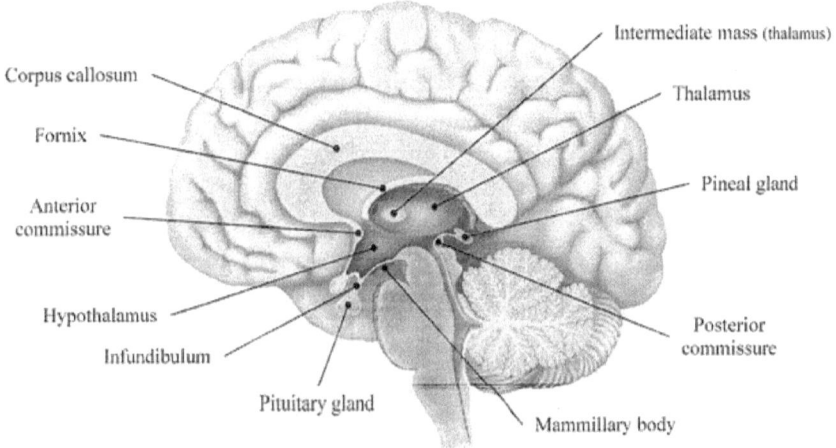

Figure 4: Mid-sagittal section of brain showing diencephalon
(includes corpus callosum, fornix, and anterior commissure)
Marieb and Hoehn
(*Human Anatomy and Physiology*, 9ᵗʰ ed.) – Figure 12.10

Exercise 2:

Utilize the model of the human brain to locate the following
structures/landmarks for the diencephalon:

- Thalamus
- Intermediate mass
- Hypothalamus
- Mammillary body
- Infundibulum
- Pituitary gland
- Pineal gland
- Posterior commissure

BRAIN STEM:

Below are listed the major anatomical regions/landmarks of the brain stem with their corresponding functions (Figure 7):

REGION / LANDMARK	FUNCTION
Midbrain	Region of brain stem between the diencephalon and pons; contains multiple fiber tracts running between higher and lower neural centers.
Cerebral peduncle	Bulge located on the ventral aspect of the midbrain; contains fiber tracts running between the cerebrum and spinal cord.

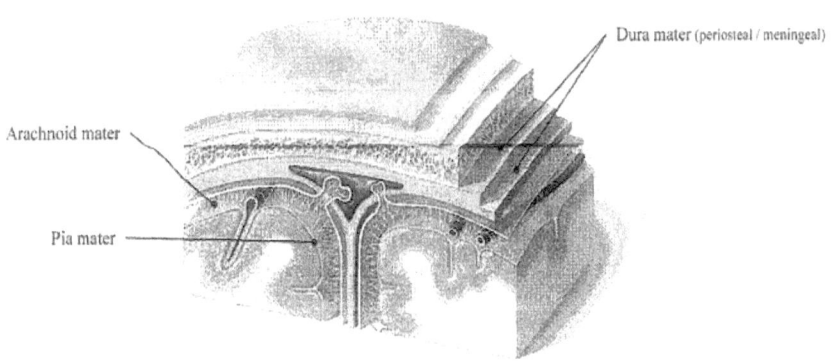

Figure 8: Section of brain and skull showing meningeal layers
Marieb and Hoehn
(*Human Anatomy and Physiology*, 9th ed.) – Figure 12.22

Knowing which part of the body is affected often allows the investigator to estimate which part of the brain is damaged before radiological procedures are performed. This is particularly true of the pre- and postcentral gyri, responsible for motor activity and sensation.

Additional workup is usually done with a CT, an MRI, or a PET scan, and usually with contrast. The brain could be roughly divided into two parts: the structures mentioned on the preceding pages form the "gray matter," the outer surface of the brain. It is convoluted and gray. Below it is the "white matter," which is made up of nerve fibers. It is like a telephone switchboard that interconnects the structures of the gray matter.

If a patient suffers from atrial fibrillation, they should take an anticoagulant called Eliquis to prevent blood clots from traveling to the brain. The chances of having a CVA (stroke) are five times higher without it.

The brain controls the ipsolateral (meaning "same side") of the body down to the level of C2. Then it switches over. The upper right brain controls the lower left body. And the lower left controls the upper right. This is a concept known as the homunculus (an imaginary upside down, back-to-front person inside the brain). It should also be noted that we perceive sensation one vertebral body above the site of stimulation. If one were to cut between C2 and C3, it would be perceived as coming from an area between C1 and C2.

Homunculus

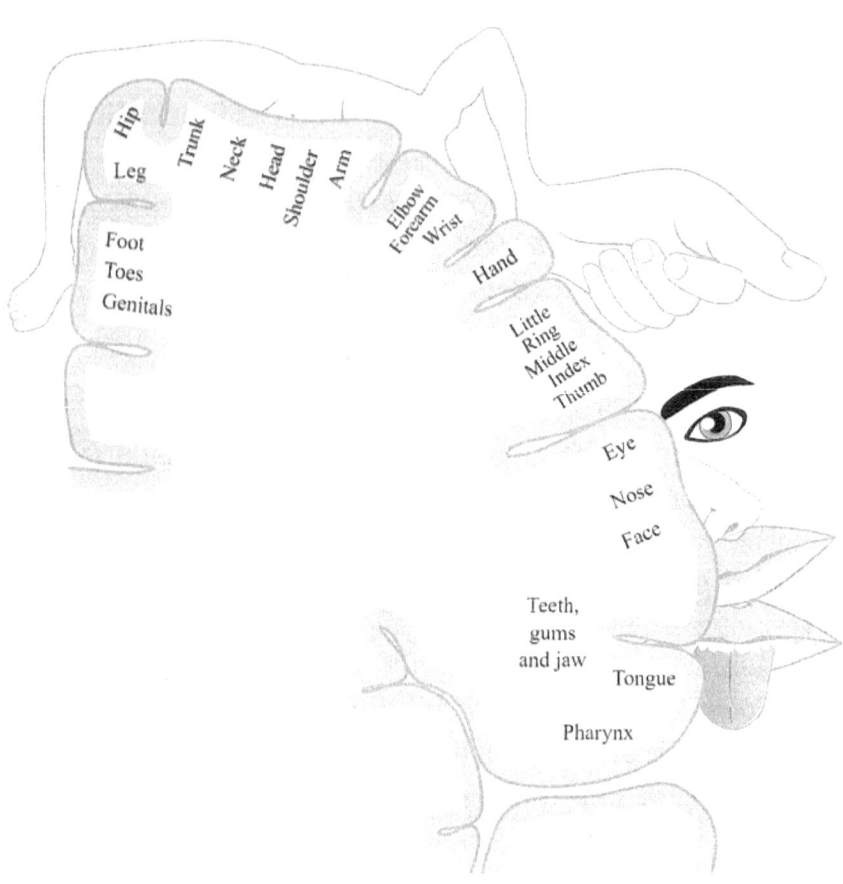

The way that the brain controls the body.

This can be used to locate neurological lesions. If there is paralysis of the left foot, you know that the lesion is in the right upper precentral gyrus. Its blood supply would, in part, be the right middle cerebral artery. If the patient presents with paralysis of the right upper arm, it tells you that the lesion is in the lower left gyrus, which receives blood from the left middle cerebral artery. Supposing the patient tells you that the paralysis is in both the left foot and the right upper arm. It is unlikely that this is being caused by a CVA. It is not likely to include blood flow to both the right upper gyrus and the left lower gyrus at the same time. But a CT or MRI would soon reveal the truth.

Posterior to the precentral gyrus is the postcentral gyrus, which is responsible for sensation and works on the same principle. It too works on the principle of the upside-down, back-to-front homunculus, and is responsible for tactile sensation. Using the concept of the homunculus, it is possible to locate most lesions before a CT or MRI is performed.

The decision to innervate a particular muscle group begins with the cerebrum. It then becomes the job of the precentral gyrus to decide which muscle groups to operate, and finally the information must go to the cerebellum, which will coordinate the muscles. Let us conclude by saying that there are complex feedback systems that tell the brain whether or not a command is correct before it is transmitted.

Once muscular information is organized by the cerebellum, it will be fed back to the cerebrum by the red nucleus before being transmitted to the cerebellar peduncles for transmission to the body. It is a way of asking the cerebrum, "Is this what you want?" before transmitting it.

Before we leave the topic of the brain, one more principle deserves discussion. How do the two halves communicate? They do so through a C-shaped structure called the corpus callosum: a neural pathway. It means "the C-shaped body" in Latin and provides communication between the two halves of the brain.

Let us move on to the spinal cord. It begins at the base of the skull and extends to L2 where it ends in the conus medullaris (meaning "middle cone" in Latin). Beyond that there are loosely packed fibers called the cauda equina that extend to L5. *Cauda equina* means "the horse's tail" in Latin.

When an organism develops, it possesses an epithelial plate. As it grows, the ends of the plate are stretched. The edges curve upward to form a neural groove and eventually a neural tube. The tube will differentiate into the spinal cord and send electrochemical messages to surrounding tissues. They will differentiate into the vertebrae that surround the cord. If the neural tube does not close, the vertebrae will not form. A condition knonw as Spina Bifida. If the failure to close affects no more than 2–3 vertebrae, it would be of no consequence and the patient, might go through life never knowing that they have it. But in its advanced state, the patient would be paralyzed from the point of opening on down.

A deficiency in selenium during fetal development (first three months) could cause this. So it is imperative for a woman to take prenatal vitamins.

The center of the spinal cord is hollow. It is surrounded by nerve fibers which make up the anterior and posterior horns of the cord.

The anterior controls muscular activity. The posterior is responsible for sensation.

On the two sides of the spinal cord (left and right), there are the spinothalamic tracts that give rise to temperature, pressure, and position sensation. The sides of the thoracic and lumbar cords give rise to the sympathetic chain ganglia. Nerve roots come out from the sides of the cord and lead to ganglia (meaning "knots" in Latin). The ganglia form a chain that runs on either side of the spine and make up the sympathetic chain ganglia.

When something touches the skin, the sensation will be perceived by the spinal root ganglia that is one vertebral process above it. For example if something were to cut the back between L2 and L3 it would be perceived at L1L2.

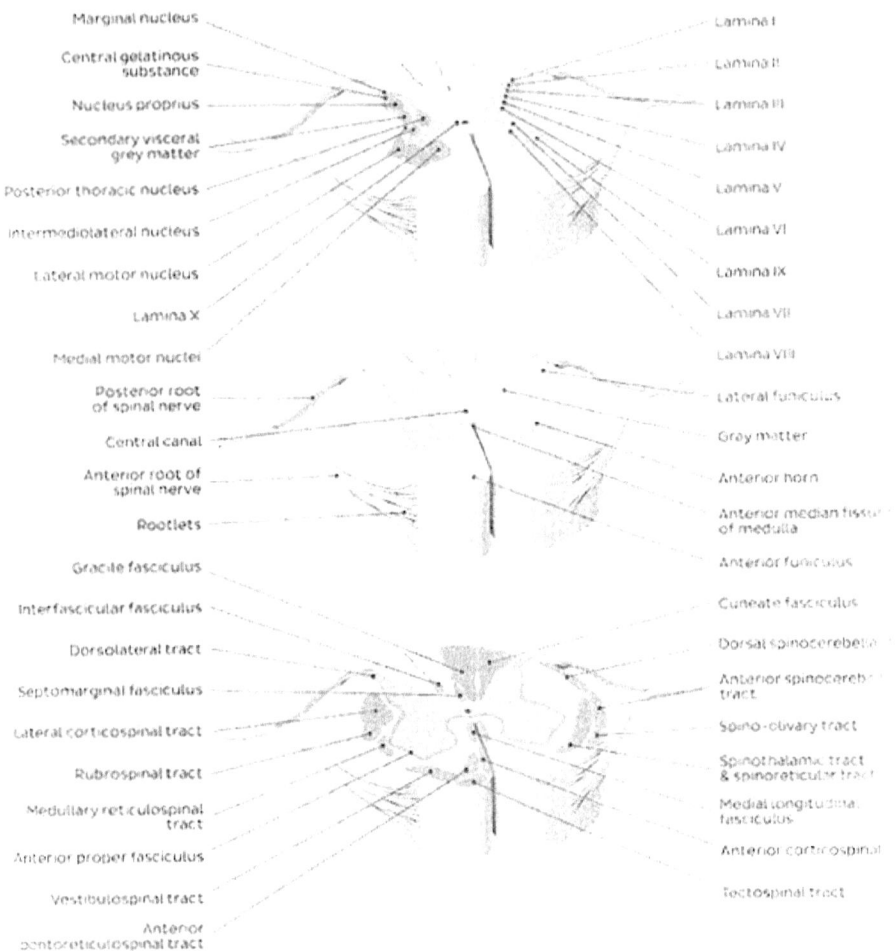

Cross-sectional view of spinal cord

Addendum

There are four conditions that affect the spine.

1. Guillain-Barré syndrome – an autoimmune disorder in which viruses change the surface antigens of neurons. This causes the immune system to attack the neurons. It leads to a breakdown of the myelin sheath that insulates the neuron, and this reduces the speed with which impulses can flow. It is treated with steroids and may or may not be reversible, depending on the extent of the damage.
2. Infections of the spine – similar to meningitis, it is often bacterial and can cripple if not treated quickly.
3. Spinal cords can have CVAs (strokes). The risk factors are be same as for the brain—smoking, diabetes, high blood pressure, and high cholesterol.
4. Spinal cords can be affected by metastatic cancer: specifically, prostate, breast, or lung melanoma

Peripheral nerves can be affected by multiple metabolic conditions such as:

Diabetic neuropathy: if fasting blood sugars are consistently above 127 and/or HgA1C is above 5.7, the neurons in the feet could be affected and lead to numbness.

Although rare in the United States, vitamin B deficiency can lead to neuropathies.

Hypothyroidism and folic acid insufficiency could affect peripheral nerves.

When patients present with neurological deficits, they need a neurological exam. Are they:

1. Alert and oriented to time, place, and person? Do they know the date, time, their location, and their identity? If not, they might be delirious.

2. Are they Romberg positive or negative? If they put their feet together and close their eyes, do they start to fall? If so, there might be a problem with the vestibular system in the ears.

3. Do they suffer from focal paralysis or numbness? If so, the pre- or postcentral gyrus might have been damaged (possibly by a CVA). An MRI would tell us for sure.

4. Do they suffer from diplopia or hemianopsia, partial blindness? If so, consider pituitary adenoma. An MRI would tell us for sure.

5. Look for hyper- or hyporeflexia. Reflexes are graded on a scale of 1 to 4 with 4 being normal. If present, consider damage to peripheral nerves.

6. Look for carpal tunnel syndrome: intermittent pain and numbness in one or both hands, accompanied by poor grip. If present, consider compression of the median nerve in the wrists. A nerve conduction velocity (NCV) and electromyogram (EMG) would tell us for sure.

7. Look for numbness of the feet. If present, consider diabetic neuropathy. A positive pinprick exam, three fasting blood sugars of 127 or higher, and a HgA1C of 5.7 or higher would confirm the diagnosis.

8. Look for cogwheel rigidity of the elbows and/or a staggered gait. If present, consider damage to the substantia nigra and Parkinson's disease.

9. Look for sudden loss of coordination and strength. Consider Guillain-Barré.

Most of the above could be worked up with a CT, MRI, PET scan, or an NCV/EMG.

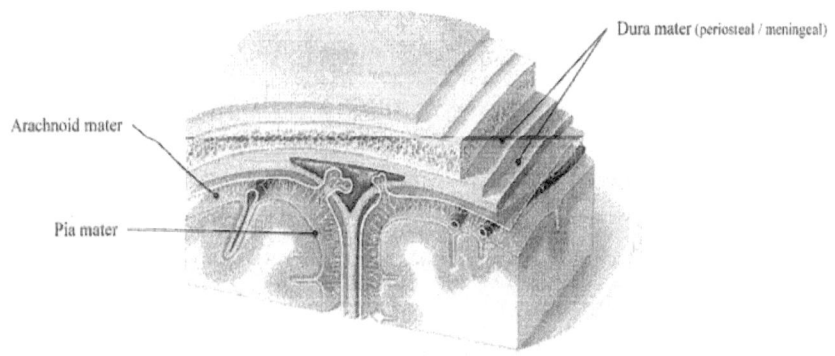

Figure 8: Section of brain and skull showing meningeal layers
Marieb & Hoehn (*Human Anatomy and Physiology*, 9th ed.) – Figure 12.22

As we said, the spinal cord is surrounded and protected by two sheets of connective tissue. The first is the pia mata mater (meaning "foot fabric tender mother" in Latin). The other is the duora matera (meaning "tough mother in Latin"). They protect and bathe the cord in cerebrospinal fluid.

The anterior and posterior horns give rise to the spinal roots, forming the spinal root ganglia, which then gives rise to nerve roots that exit the vertebrae through the vertebral foramen. From there the nerves circle the body through the skin dermatomes or with the ribs.

From this point on the nerve roots give rise to the systemic nerves. The best way to learn them is to memorize mnemonics.

The head is innervated by 12 cranial nerves:

> First: optic nerve (vision)
> Second: olfactory nerve (sense of smell)
> Third: oculomotor nerve (pupil contraction)
> Fourth: median trochlear nerve (medial gaze)

- Fifth: trigeminal nerve originates from the semilunar ganglion anterior to the ear and forms three branches: temporal, which innervates the face above the eyes; maxillary, which innervates the face between the eyes and the mouth; and mandibular, which innerovates the lower face). It provides the face with tactile sensation.

> Sixth: lateral trochlear nerves responsible for lateral gaze.
> Seventh: facial nerve. Provides the face with muscular innervation and has five branches:

Temporal – innervates the upper part of the face.

- Zygomatic – innervates the face over the zygomatic arch.
- Buccal – innervates the cheeks.
- Mandibular – innervates the face over the mandible.
- Cervical – innervates the upper neck,

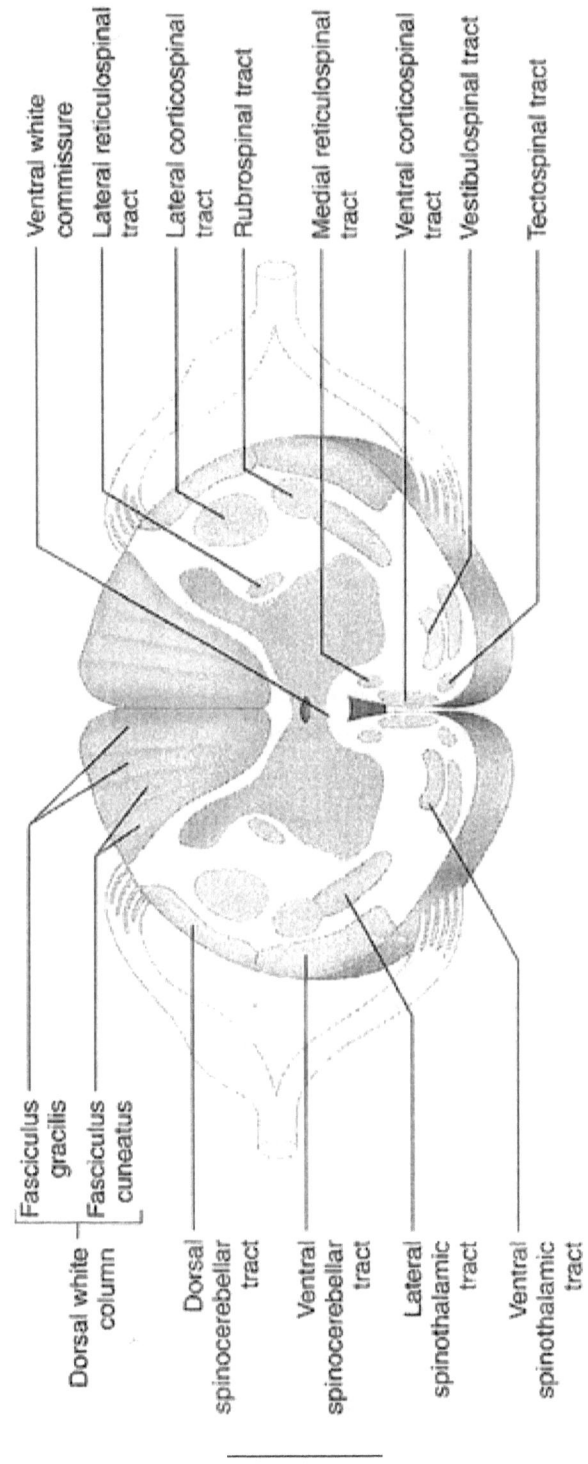

Seventh cranial nerve (facial nerve) mnemonic:

Zygomatic innovates the face over the zygomatic arch

To	Temporal
Zanzibar	Zygomatic
By	Buccal
Motor	Mandibular
Car	Cervical

- The eighth cranial nerve is the vestibular-ocochlear. One branch innerovates the vestibular canals (also known as the semicircular canals). There are three of them, and they tell the brain where the body is (positioning sense). One canal lies in a coronal plane (separating the body into anterior and posterior). The second lies in a transverse plane (dividing the body into inferior and superior), and the third lies in a sagittal plane (that divides the body into left and right).

Semicircular canals:

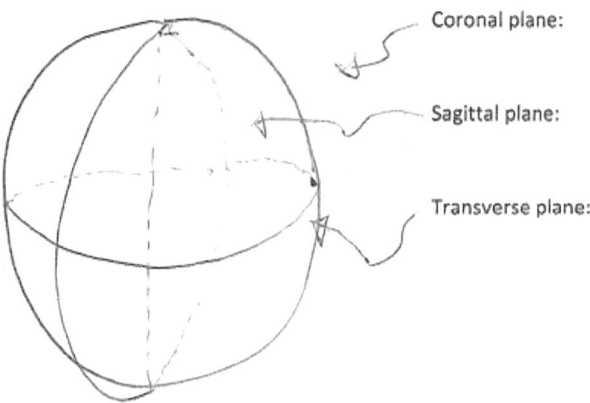

Coronal plane:

Sagittal plane:

Transverse plane:

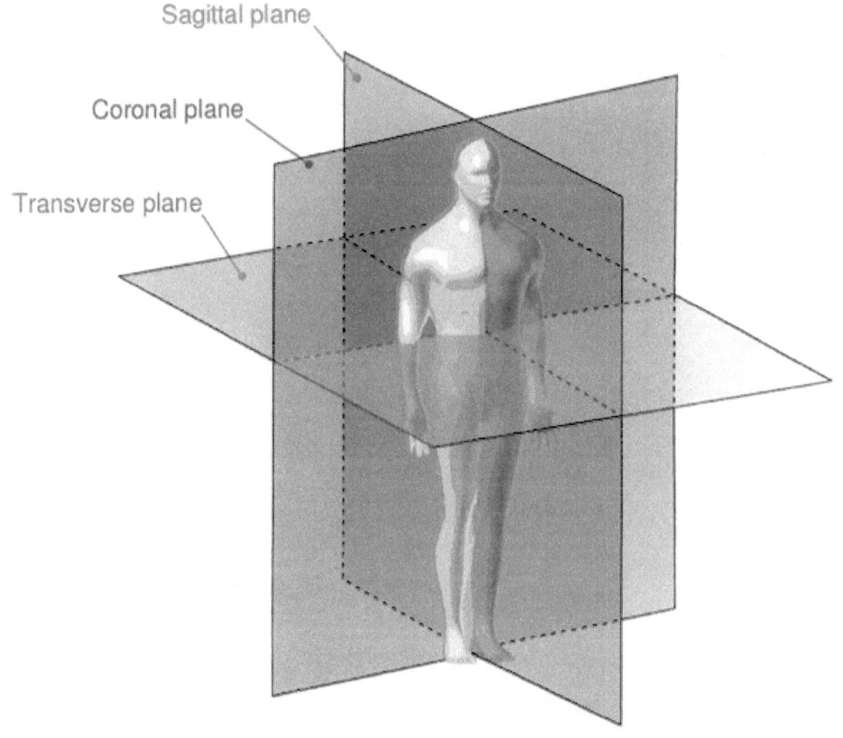

Sagittal plane

Coronal plane

Transverse plane

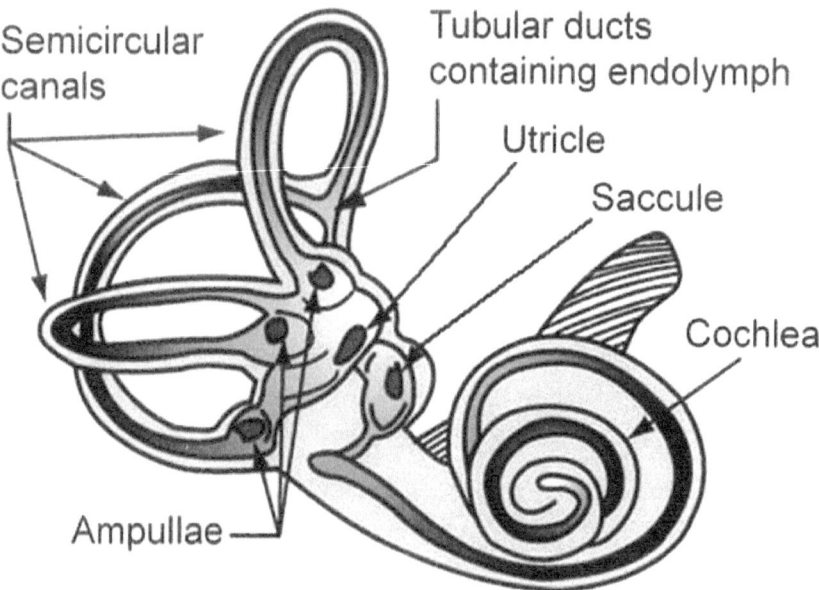

Figure 2: The Vestibular System - semicircular canals and otolith organs

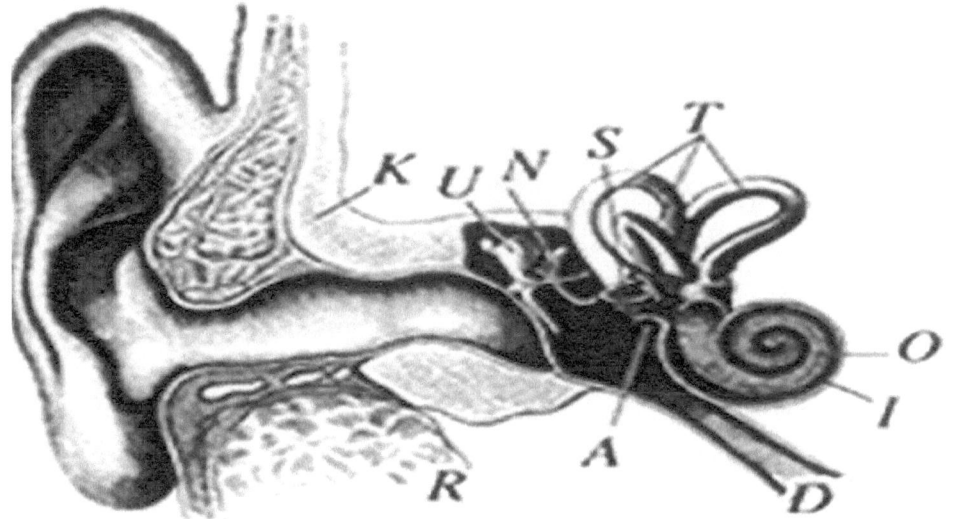

If the canals become inflamed (as with Ménière's disease: too much salt), the patient presents with vertigo (dizziness) and they would be Romberg positive (if they put their feet together and close their eyes, they will start to fall). This can sometimes be treated with meclizine (25 mg by mouth three times daily or as needed) and/or salt restriction.

The second branch of the vestibulocochlear nerve goes to the cochlea (which means "snail-like" in Latin). It makes up the inner ear. It looks like a snail, is fluid-filled, contains ciliary fibers, and converts vibration into neural impulses. The eighth cranial nerve enters the brain through the brain stem.

- The Ninth cranial nerve is the glossopharyngeal, that which extends the tongue and plays a role in taste appreciation.
- Tenth cranial nerve is the vagus. It is a parasympathetic nerve that provides parasympathetic innervation between the head and the middle of the transverse colon (upper body). The lower half (transverse colon to feet) is innervated by parasympathetic nerve fibers that arise from the sacral vertebrae.
- The eleventh cranial nerve is the spinal accessory that shrugs the shoulders. It innervates the sternocleidomastoids and the trapezius muscles.
- Finally, the hypoglossal (cranial nerve number twelve) extends the tongue.

The cranial nerves originate from cranial nuclei on the same side of the brain. Once we go below C2 (second cervical vertebra), innervation switches over, and the brain controls the opposite side of the body.

As we drop below C2, we have to think of neurological innervation in terms of the "homunculus," an imaginary upside-down, back-to-front person within the brain. The right side of the body is controlled by the left side of the brain. The lower body is controlled by the upper parts of the brain, and vice versa.

Addendum

Let us talk further about the peripheral nerves.

The brachial plexus – This is a group of nerve cords in the axillae (armpits). They arise from cervical vertebrae 2–6 (C2–C6) in the neck. They give rise to the median, ulnar, and radial nerves in the arms. These nerves innervate the arms and hands.

Phrenic nerves arise from C3, 4, and 5 on either side of the neck and innervate the diaphragm. Remember, "3, 4, and 5 keeps the diaphragm alive."

The intercostal nerves innervate the intercostal muscles. Remember, "VAN" (vein, artery, nerve).

The parasympathetic nerves that arise from sacral vertebrae 2, 3, and 4 cause the penis to erect.

Remember, "2, 3, and 4 keeps the penis off the floor."

The sciatic nerves, which innervate the legs and the feet, arise from the lumbar vertebrae. Sciatica is a condition in which the patient complains of pain radiating down one or both legs. It is caused by compression of the nerve roots as they exit the neural foramen.

Let us consider infections of the brain.

Meningitis

- Herpes encephalitis

Cysticercus (rare in the United States)

Meningitis is most prevalent in adolescents and young adults and can be avoided by giving meningococcal vaccine. Once they have it, it must be treated emergently with IV (and possible intracranial) antibiotics such as Ampicillin and Cefotaxime or Ampicillin and Gentomicin (The Sanford Guide to Antimicrobial Therapy, 2008 pg. 6).

It typically presents with:

- Severe headaches

 Altered consciousness
 Nuchal rigidity (stiff neck)

It can kill or disable in less than a day.

Herpes encephalitis is most common in immunocompromised patients and should be suspected in someone who develops sudden choreiform (undulating) movements of the head. It should be treated with intrathecal medication (such as acyclovir).

Cysticercus is the larval form of a tapeworm and might present as a new-onset seizure in a person who ate contaminated food (especially rare fish). It develops in the spinal region of the fish. It would be diagnosed with an MRI. Its treatment is beyond the scope of this publication, but the patient would have to be treated with an anthelmintic (medication for parasitic worms). This would have to be done slowly, because if the worm dies too quickly, its decomposition products may further damage the brain. An infectious disease (ID) specialist should be involved.

In the 1960s the world faced a threat that could have annihilated the human race: biological weapons. An international conference was held at Versailles near Paris to outlaw them. Richard Nixon was in attendance. These weapons were so deadly that one attack could have wiped out the Eastern Sea Board of the United States.

There are six principles:

1
Start with bacteria and viruses that are deadly to begin with such as The Spanish Influenza, Smallpox, Yellow fever, and plague.

2
Treat them with multiple antibiotics, antivirals, and vaccines and cultivate the ones that are immune.

3
Mix several deadly organisms. They will not simply add to each others effects. They will multiply them. So, let us say that we have two organisms that cause a loss in blood pressure. If we combine them they will not just add to each others effects. They will multiple each others effects and affect blood pressure much more than either one alone. This is called 'synergy.'

4
Mix two or more organisms in which the antidote for one will make the other worse.

5

Take a deadly virus or a prion (which can not be cured) but is not very contagious. Something like e-bole or kuru, and allow it to infect a microplasm: a very small bacteria that is barely visible under a light microscope. Microsplasms are contagious, hard to filter out, and can travel long distances in air.

When the victim ingests them they would think they had flu. It would pass in a few days. But when the immune system destroys the microplasm they would release the virus or prion which would cause death.

6

The deadliest known out brake occurred in 1918. It was the Spanish Influenza and killed 120 million people. Almost twice the number who died in World War II (68 million). It was a flu that affected multiple species. If it infected a dog it would take some of its DNA. Then it would infect a human and place the canine DNA into a person. This would change the antigens on the patients organs and the organs would be attacked by their own immune system. It is also difficult to vaccinate against flu because its surface antigens change often.

Musculoskeletal System

Let us move on to the musculoskeletal system. As said, there are about 600 muscles and 206 bones in the adult body. During embryology they arise from mesoderm (meaning "middle skin").

Muscles serve four functions:

Muscles allow us to move. They also move air, pump blood, and make it possible for us to digest food.

As we said there are three kinds of muscle:

- Cardiac, which pumps blood

  Smooth, which is responsible for involuntary movements
  Striated, which makes voluntary movements

Bones are made of hydroxyapatite (a form of calcium phosphate, which is water-insoluble), and there are two kinds:

- Long bones, which support the body

  Flat bones, which contain the bone marrow. They produce red blood cells (that transport oxygen and carbon dioxide), white blood cells (responsible for immunity), and platelets (responsible for clotting).

Bones also provide protection, and the smallest bones of the body (the anvil, stapes, and malleus) make it possible for us to hear.

If we make a transverse cut across a long bone, we will see the Haversian system (that looks like a tree trunk with rings and knots). At the very center is the bone marrow, and the rings tell us how old the bone is. The knots are blood vessels. If we made a longitudinal cut we would see what looks like a honeycomb. This is a lightweight structure that gives bone tensile strength.

Let us make a running list of the different bones of the body.

The skull consists of the:

- Parietal bone

    Occipital bone
    Temporal bone

    Frontalis bone
    Zygomatic arch
    Maxilla
    Mandible
    Hyoid bone
    Petrous ridge
    Sphenoid bone

The spine in the neck consists of seven vertebrae. Each vertebra consists of:

- Anterior body

    Two lateral spines (processes)
    One-posterior process
    Two vertebral foramens, allowing the nerve roots to exit

The thoracic spine lies behind (or posterior to) the chest and consists of 12 vertebrae.

The lumbar spine is posterior to the abdomen and contains 5 vertebrae.

The sacrum is a fused spine and consists of 5five fused vertebrae that lie behind the pelvis.

Extending from each thoracic vertebra is a rib. The first eight join the sternum in the middle of the chest. The sternum can be further divided into:

- Manubrium

    Body
    Xiphoid process

At the very top of the thorax (or chest) are the clavicles (collar bones), which attach to the manubrium medially and to the acromioclavicular processes of the scapulae laterally.

Posterior to each shoulder lies the scapula or shoulder blade. These are wing-shaped bones on either side of the upper back.

In the pelvis we have the iliac crests that provide support and protection and contain biologically active marrow in adults (red marrow).

Between the iliac crests and in the posterior region of the pelvis is the sacrum, which is connected to the left and right iliac crests by the sacroiliac joints. This is a common place for arthritis to develop.

In the anterior part of the pelvis is the pubic symphysis, and in the inferior parts of the pelvis we find the femoral acetabulums (sockets of the leg joints). The inferior part of the pelvis is bounded by the floor of the peritoneal cavity and the ischial tuberosities (which is what we sit on).

The thigh bones are the left and right femurs, and the kneecaps are the patellas. The proximal end (closest to the body) of the femur is the femoral head, which inserts into the femoral acetabulum. It is attached to the femoral neck, which extends laterally and forms the greater trochanter.

The distal ends of the femurs form the condyles, which rock back and forth on the medial and lateral meniscus and on the plateau of the tibia (the larger bone of the lower leg).

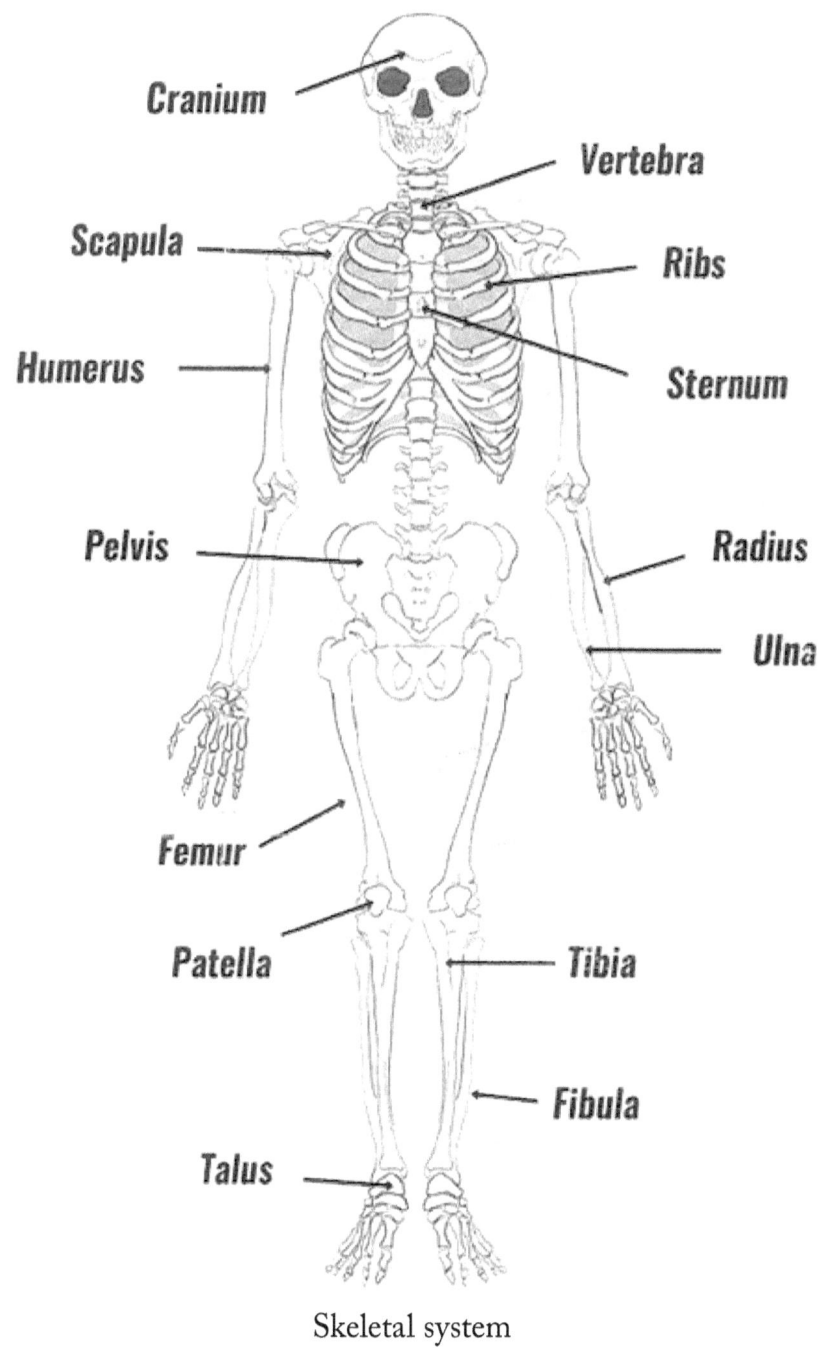

Cranium

Vertebra

Scapula

Ribs

Humerus

Sternum

Pelvis

Radius

Ulna

Femur

Patella

Tibia

Fibula

Talus

Skeletal system

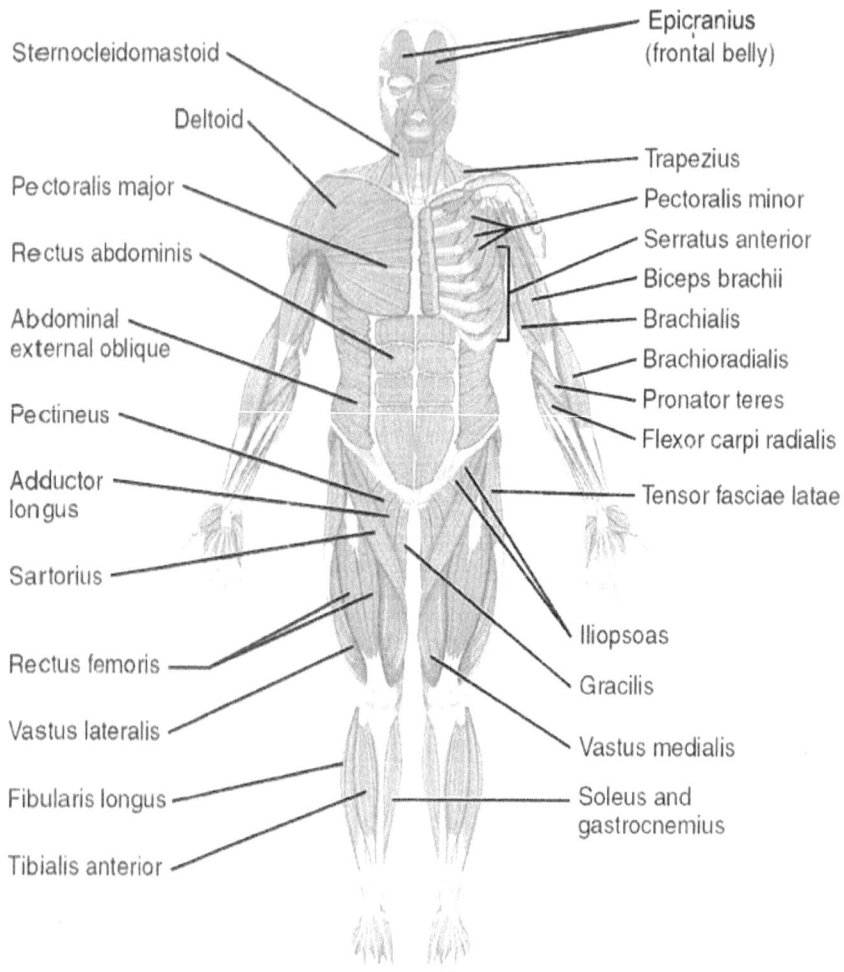

Sternocleidomastoid

Deltoid

Pectoralis major

Rectus abdominis

Abdominal
external oblique

Pectineus

Adductor
longus

Sartorius

Rectus femoris

Vastus lateralis

Fibularis longus

Tibialis anterior

Epicranius
(frontal belly)

Trapezius

Pectoralis minor

Serratus anterior

Biceps brachii

Brachialis

Brachioradialis

Pronator teres

Flexor carpi radialis

Tensor fasciae latae

Iliopsoas

Gracilis

Vastus medialis

Soleus and
gastrocnemius

Major muscles of the body.
Right side: superficial; left side:
deep (anterior view)

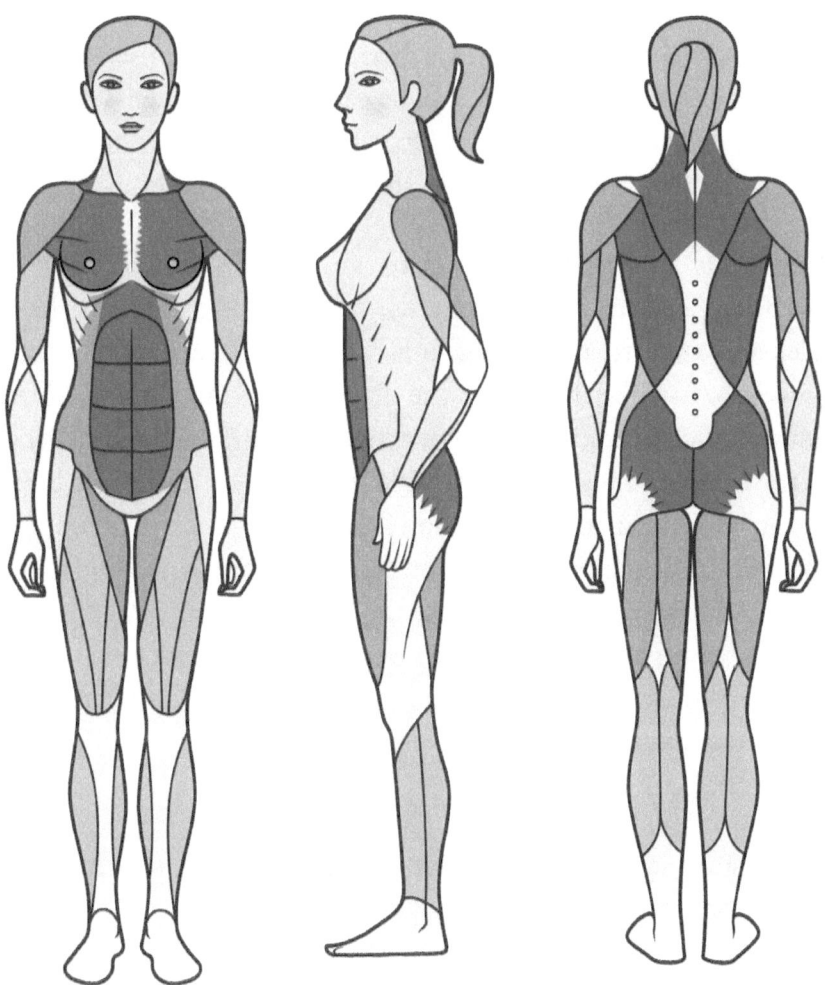

Female Muscular System

The tibias are the shin bones and are accompanied by the fibulas (smaller bones in the lower posterior legs). If a fibula were to break, the patient could still walk, but not if a tibia breaks.

The distal ends of the tibias form a concave joint that fits onto the top of the talar domes, and the talar domes sit atop the left and right talus (the ankle bone). Each Achilles tendon inserts on the posterior aspect of the talus, a bone called the calcaneus. Anterior to the talus are the tarsals (the foot bones), and anterior to those are the metatarsals. Then come the toes or phalanges.

Now let us consider the anatomy of the arms and hands. The bone in the upper arm is the humerus, which, like the femur, has a neck and a greater trochanter. It is the origin of several upper arm muscles. And like the femur, the distal end of the humerus forms double notched epicondyles that articulate with the proximal ends of the ulna and the radius. The ulna is the thicker of the two bones in the lower arm, and its position is relatively fixed. The radius is the smaller of the two, and it rotates around the ulna. The radius allows us to supinate and pronate the hand. When we supinate our hands, we turn them upward so that they could hold soup. If we pronate them, we turn them the other way, so that they face down. There is a muscle distal to the elbow that rotates the radius. It is called the pronator teres.

The distal end of the radius forms the styloid process, a fingerlike projection.

Distal to the radius and ulna are the carpal bones (wrist bones):

- Scaphoid

- Triquetrum

  Capitate

- Trapezoid

  Trapezius
  Hamate
  Lunate

- Pisiform

Distal to that are the five metacarpal bones of the hand, and distal to that are the finger bones or phalanges. The thumb has a proximal and a distal phalanx and the other fingers have a proximal, middle, and distal phalanx.

There are three other bones that deserve mention:

- Incus (anvil)

  Stapes (stirrup)
  Malleus (hammer)

These are the smallest bones in the body, and they are susceptible to arthritis, infection, and trauma. They are the bones of the middle ear and are essential for hearing. When sound waves enter the ears, they do so through the external auditory canals. Then they vibrate the tympanic membranes (often referred to as the eardrums). The tympanic membranes then move the incus, stapes, and malleus. They vibrate the window of the cochlea, which converts the vibration to electrical impulses.

If an injury to the middle ear is suspected, the investigator can perform a Rinne and Weber test. They start with a conventional hearing test. If it reveals diminished hearing in one of the ears, they place a tuning fork on the center of the forehead and ask the patient which side sounds louder. If the sound is loudest in the affected area, that indicates that the bones of the middle ear have been damaged. When they place the fork on the forehead, the patient is hearing by way of bone conduction. The sound is traveling through the bones of the skull directly to the cochlea. If the bones of the middle ear are not functional, the cochlea will have adjusted itself to be more sensitive, and the patient will perceive the sound as being louder in the affected ear. A subsequent MRI will reveal exactly where the problem lies.

Let us talk about the muscular system. For the sake of brevity we will discuss only the major muscles. There are several big muscle groups in the head. They start with the temporalis, which are close to the temples. When temporalis muscles become tense for a prolonged period, they create tension headaches. Their normal role is to give expression to the upper face, but if they remain tense, they create pain, and it can be severe. Rest, hydration, and simple medication such as Tylenol or ibuprofen are all that is usually needed.

The zygomatic muscles lie over the zygomatic arch. Below that are the buccal muscles which move the cheeks. Then there are the mandibular muscles which overlie the lower jaw, and finally, the cervical muscles which move the upper neck.

The eyes are closed by the orbicularis oculi (circular muscles that surround the eyes) and the mouth is closed by the orbicularis oris. There are the muscles of the tongue that we already discussed.

There are two major muscle groups in the neck:

- The sternocleidomastoids (one on the right and one on the left).

    The trapezius muscle in the upper back and in the neck.

The sternocleidomastoids originate at the junction of the collarbone and the manubrium and inserts on the mastoid air cells on both sides of the skull. They are responsible for nodding the head. The trapezius shrugs the shoulders. The trapezius originates along the borders of the upper thoracic and cervical spines and inserts at the base of the skull.

The trapezius is a trapezoid shaped muscle that originates along the thoracic and cervical vertebrae. Its lateral boarders extend to the spines of the scapula (left and right) and its upper board inserts at the base of the skull.

It assists in posture and movement of the neck. It also assists in rotation of the arms and in shrugging the shoulders. It is innervated by the XI cranial nerve (accessory) and receives its blood supply from the thoracic aorta and the arteries that accompany the ribs (VAN).

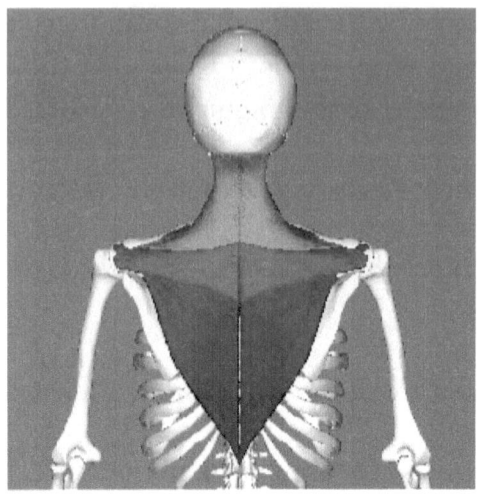

In the larynx we also have the vocalis muscles that move the vocal cords and the arytenoid cartilages. They are innervated by the recurrent laryngeal nerves (one on the left and one on the right). There is a muscle that closes the epiglottis during swallowing so that food will not enter the lungs.

The muscles of the shoulder and the arm include the rotator cuff, which helps to turn the arms laterally and medially and is innervated by branches of the brachial plexus. In the upper arm we have the triceps, which extend the elbow, and the biceps, which flex it. In the lower arm we have the pronator teres which turns the radius relative to the ulna and is innervated by a branch of the median nerve. It receives blood from a branch of the brachial artery.

In the lower arm there are two sets of muscles that open and close the hands: the palmaris brevis and longus in the posterior compartment, which open the hands, and the palmaris brevis and longus in the anterior compartment, which close them. They are innervated by the ulnar and radial nerves and receive blood from branches of the brachial arteries. They insert on ligaments that go under the retinacular ligament in the wrists and into the hands.

In the hands we have the interosseous muscles that are responsible for fine movements. They receive blood from branches of the palmar arches and neurological innervation from branches of the radial and median nerves. They are also called the lumbricals, which means "worms" in Latin.

Moving down into the chest, we find several large muscle groups:

- Pectoralis minor

  Pectoralis major

  Intercostal muscles

The pectoralis major and minor muscles are innervated by the median and lateral pectoralis nerves that originate in the c-spine. They receive blood from arteries and veins that course with the ribs. They are responsible for abducting the arms (pulling them inward). The intercostal muscles receive their blood supply and neurological innervation from the same nerves and vessels. They pull the ribs together (VAN). This assists in respiration.

The muscles of the upper back include deep back muscles and the latissimus dorsi. They are also innervated by intercostal nerves and by the thoracodorsal nerve (from the brachial plexus). The latissimus dorsi pulls the arms back, and the deep back muscles assist with posture.

Besides the stomach and intestinal wall, the abdomen contains one major muscle: the rectus abdominus. This is a longitudinal muscle that runs down the center of the abdomen and tenses in response to pain. The abdominal wall also contains muscle fibers.

In the thigh there are several muscle groups such as the hamstrings which flex (draw up) the lower leg.

The vastus muscles (the quadriceps) in the anterior compartment of the thigh extend the lower leg. Below the knee are the calf muscles. In the posterior compartment of the lower leg is the gastrocnemius, which pulls the Achilles tendon and extends the foot. Within the foot there are lumbricals that assist in fine movement.

Muscles of the upper and lower leg and foot are innervated by branches of the sciatic nerves. Their blood supply comes from branches of the femoral and popliteal arteries.

The Endocrine System

Let us move on to the study of glands. There are two kinds:

- Endocrine

  Exocrine

Endocrine glands secrete into the bloodstream and include:

- Pineal

  Pituitary
  Thyroid
  Pancreas
  Adrenals
  Reproductive organs (testicles, ovaries)

The endocrine system is key to our body's chemistry.

The pineal gland is a biological clock and sets the natural rhythm for our bodies known as the circadian rhythm. This is a 24-hour rhythm. It peaks at about noon (12 p.m.), falls to intermediate levels at around 6 p.m., reaches its low point between 12 midnight and 2 a.m., and returns to an intermediate level at around 6 or 7 a.m.

The pineal gland also secretes melatonin, which assists with sleep. It controls our rate of metabolism, so we metabolize the least at midnight and the most at noon.

The pituitary gland, with an anterior and posterior lobe, secretes many hormones including follicle stimulating hormone, luteinizing hormone, thyroid stimulating hormone, and growth hormone.

The thyroid gland secretes T3, T4, and T7, which drive the metabolism of our bodies.

The parathyroid glands regulate the concentration of calcium in blood and in bones.

The thymus gland in our mediastinum (center of the chest) plays a key role in immunity.

The adrenal glands lie above the kidneys and supply:

- Adrenalin (for fight or flight)

    Glucocorticoids: reduce inflammation and raise blood sugar
    Mineralocorticoid to retain water
    Small doses of estrogen
    Small doses of testosterone

We also have the ovaries and testicles, which secret estrogen and testosterone, and the kidneys. Kidneys produce important substances such as erythropoietin, which stimulates bone marrow to synthesize red blood cells, and angiotensinogen, which raises blood pressure.

The pancreas is an endocrine and an exocrine gland. It secretes digestive juices such as lipase, into the GI tract. It also releases insulin into the circulatory system.

Exocrine glands secrete into hollow cavities and outside the body. They include:

1. Wax secreting glands within the ears
2. Salivary glands
3. Mucous glands of the sinuses and bronchial passages
4. Digestive glands within the GI tract
5. Sweat glands of the skin
6. Mammary glands, which secrete milk soon after birth
7. Pheromones: hormones that act outside the body. They typically tell other members of the species that a female is ready to reproduce.

When the body needs a particular substance, it will send a chemical messenger to the glands. One of them will contain a segment of DNA that will open and expose the "sense strand." This will code for mRNA that will leave the nucleus and enter the RER. This will lead to a cascade of reactions that will cause a particular substance to form. The substances will be packaged into a vacuole or Golgi body and will be secreted through reverse pinocytosis or phagocytosis.

Let us look at the mechanism by which glands synthesize and secrete special substances, and let's discuss the mammary gland. The cell membrane expels substances into the environment that surrounds the cell. This process is known as reverse phagocytosis and is the basis of cells secreting solids (in the form of a colloid such as milk) and liquids. This is how glandular tissues work. They secrete liquids and solids into the surrounding environment through the above mechanism. The process of expelling liquids is actually referred to as reverse pinocytosis although the mechanism is roughly the same.

Lipids in the form of liquids and solids are synthesized in the SER. They are then packaged into Golgi bodies that deliver them to the cell membrane. Then they are expelled into the surrounding environment through reverse pinocytosis and phagocytosis.

The RER does the same with liquid and solid proteins, and collectively this accounts for the endocrine and exocrine glands' ability to secrete useful substances: A classic example is the mammary gland, which secretes milk soon after birth.

If you look at a breast on a mammogram, you see thousands of tunnels (known as ducts) that terminate at the nipple. If you examine these ducts under a microscope, you see a tunnel known as a lumen surrounded by bottleneck cells that look like kernels of corn. If you were to stain the cells with shifts stain, we would see thousands of small particles within the cytosol that would have a violet color. These are the vacuoles and Golgi bodies that have incorporated proteins and lipids made in the rough endoplasmic reticulum and in the smooth endoplasmic reticulum.

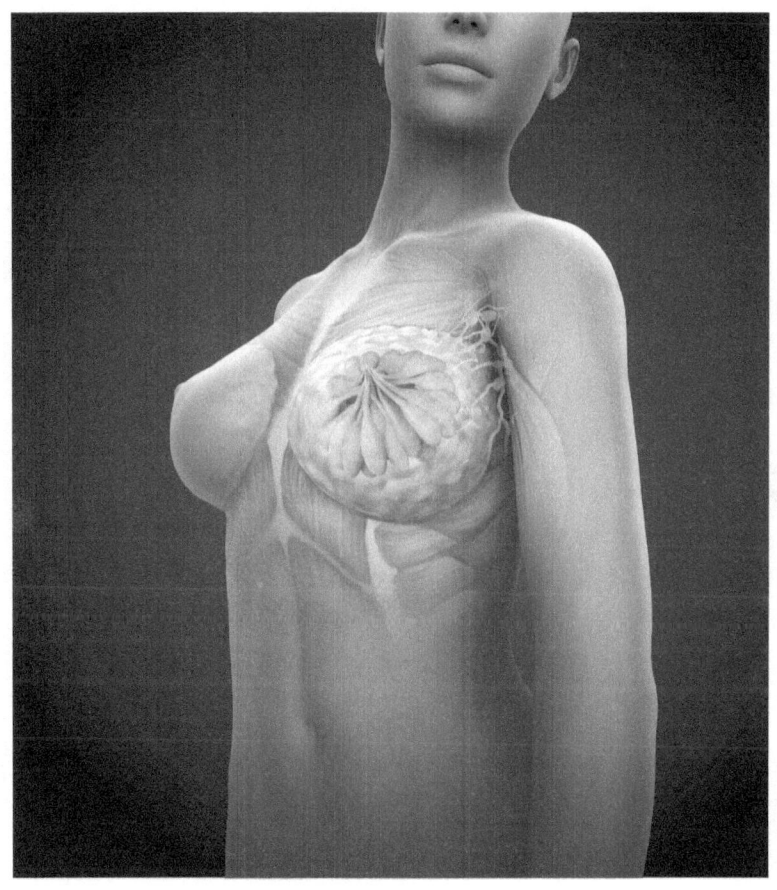

Mammary glands

Mammary glands

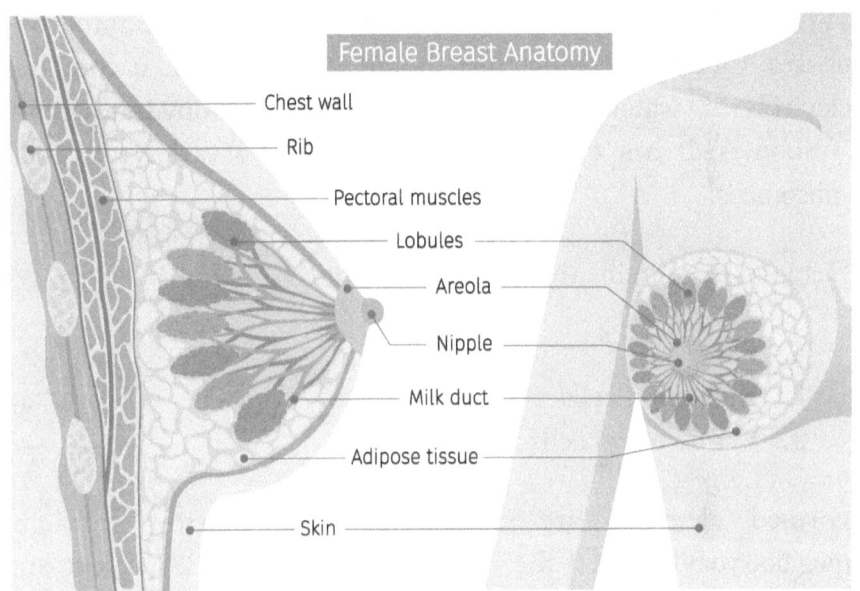

Female Breast Anatomy

Chest wall

Rib

Pectoral muscles

Lobules

Areola

Nipple

Milk duct

Adipose tissue

Skin

Mammary glands

Addendum

Hematogenous System

Let us think about the only liquid tissue in the body: blood. An adult has five liters and it has two components:

1. Hematocrit
2. Plasma

When tubes of blood are centrifuged, they separate into two phases. One is a straw-colored liquid called plasma. It contains electrolytes, water-soluble proteins, lipids, carbohydrates, and antibodies. It has a pH of 7.4 and contains a variety of other substances.

Hematocrit contains the formed elements:

1. Platelets
2. Red blood cells
3. White blood cells

Formed elements are created in bone marrow through megakaryocytosis. Of the cells in bone marrow, 98 percent are committed cells and do not reproduce. But 2 percent are stem cells, which are totipotent and create the formed elements. Most red marrow eventually becomes yellow marrow (between the ages of thirteen and seventeen) with a high concentration of lipids. It will be physiologically inactive.

Adults typically have red marrow in two locations:

1. The sternum (center of chest)
2. Iliac crests of the pelvis

Under conditions of extreme oxygen deprivation (such as high altitudes), yellow marrow can revert back to red marrow. But this is rare. If doctors sample the red marrow, they get it from the iliacs with a needle called a trochar.

Red blood cells transport oxygen and carbon dioxide. They have the ideal shape for absorbing the greatest amount of gas in the shortest period of time, and they can release oxygen with greater ease as pH drops.

Red blood cells:

- Last 120 days

  Do not contain a nucleus

  Contain a protein called hemoglobin, which becomes redder as it absorbs oxygen

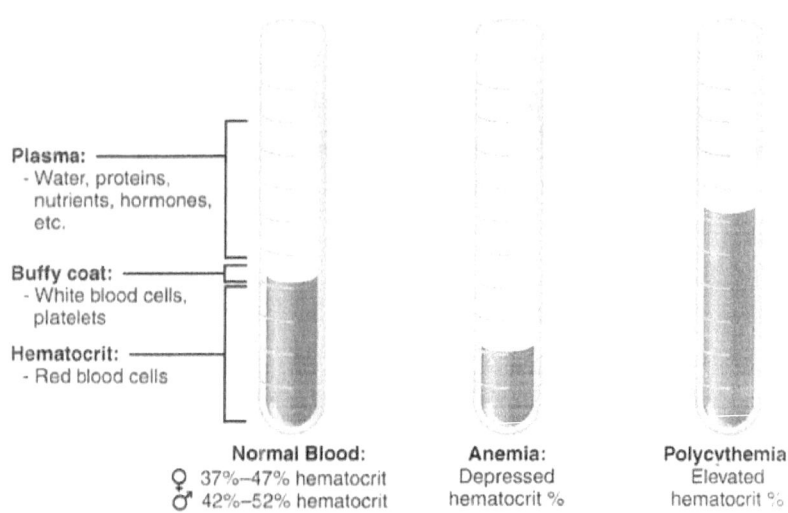

Plasma:
- Water, proteins, nutrients, hormones, etc.

Buffy coat:
- White blood cells, platelets

Hematocrit:
- Red blood cells

Normal Blood:
♀ 37%–47% hematocrit
♂ 42%–52% hematocrit

Anemia:
Depressed hematocrit %

Polycythemia:
Elevated hematocrit %

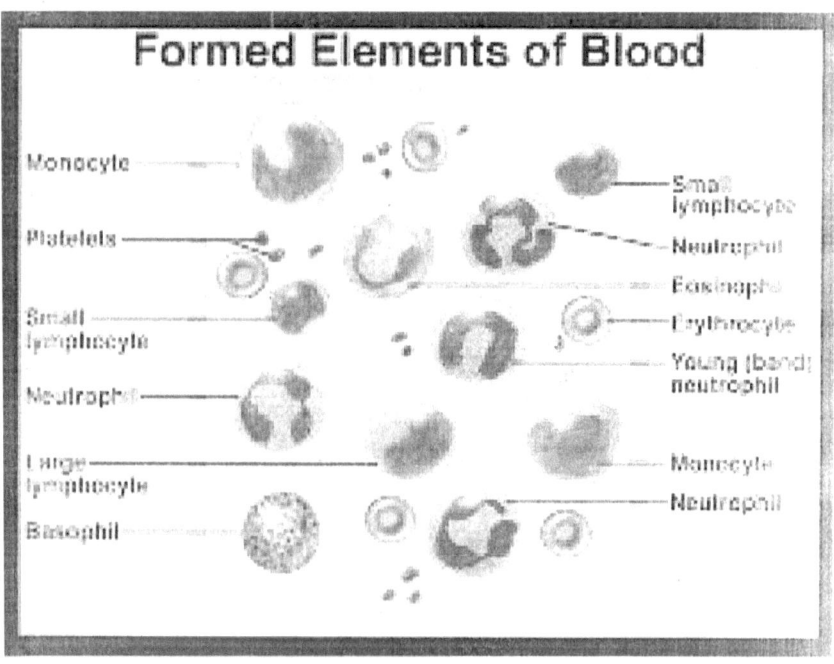

Formed Elements of Blood

Monocyte

Platelets

Small lymphocyte

Neutrophil

Large lymphocyte

Basophil

Small lymphocyte

Neutrophil

Eosinophil

Erythrocyte

Young (band) neutrophil

Monocyte

Neutrophil

Blood: A Liquid Tissue

Occasionally a condition known as polycythemia rubra vera will cause the bone marrow to make too many red blood cells, and it will be necessary to hemorrhage the patient from time to time.

White blood cells fight infections, consume dead cells, and suppress cancer. They do it in three ways that we will expand upon later.

Platelets are responsible for coagulation. We will discuss this in more detail later. But they too do not have a nucleus and last for seven days.

The plasma provides:

• Nutrition

Immunity (with water-soluble antibodies)

pH regulation

Elimination of wastes

Delivering hormones to different parts of the body

Two percent of the cells in bone marrow are stem cells. They are totipotent meaning that they could become any cell in the body. But they become the formed elements in accordance with the diagram that appears below.

Totipotent Bone Marrow Cells
(formed elements of the blood)

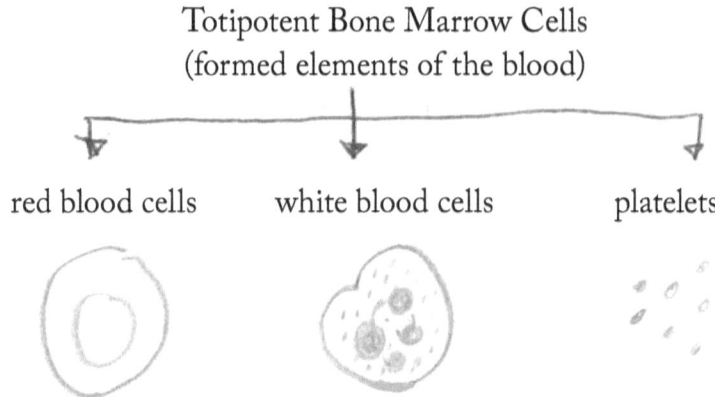

red blood cells   white blood cells   platelets

The ability of white blood cells to fight infections is referred to as the "humoral" response.

Some white blood cells such as neurtrophils, eosinophils, monocytes, lymphocytes, and mast cells attack foreign invaders and cancers directly. Others such as T-helper cells package the antigens of invaders and transport them to a gland in the mediastinum called the thymus. They are presented to T-killer cells that attack the invader. And finally there are lymphocytes that become B cells which differentiate into plasma cells. They have a rich plexus of RER and synthesize antibodies, especially IgG and IgM.

IgM is a larger molecule than IgG and cannot cross the placenta. But IgG can.

Blood: A Liquid Tissue

Plasma cells with a rich plexus of RER
(which makes antibodies)

Now let us talk about different blood types. Blood is characterized as A, B, AB, or O and as Rh⁺ or Rh⁻. This system was devised in the 1930s and refers to antigens (proteins) on the surface of red blood cells.

If we need a transfusion we could determine the type we are going to need in accordance with the following:

> If we have type A, we may receive type A or type O blood.

> If we have type B, we may receive type B or type O blood.

> If we have type O, we may only receive type O blood.

> If we have type AB, we may receive type A, B, AB, or O blood.

So type AB is the universal recipient, and type O is the universal donor.

> If we have type AB, we may receive blood from anyone.

> If we have type O, we may give blood to anyone.

Here's how it works. Our body's immune system will not react to an antigen that we are born with. So, if we have type A antigens on our red cells, our immunity will not react to it. But if we are exposed to A-like proteins in the form of foreign invaders and if we are not naturally A, our immunity would react to it. The same would be true of B. O is the absence of an antigen. It means that our cells have no antigen at all.

Their production is stimulated by a hormone, called erythropoietin, that is released by the kidneys.

Clinicians classify blood in terms of its ABO status:

    a. If it contains no surface antigens, it is type O blood.
    b. If it contains A antigens, it is type A.
    c. If it contains B antigens, it is B.
    d. If it has A and B antigens, it is type AB.

Patients with type O are the universal donors and can give blood to anyone. Those with type AB are the universal recipients and can receive blood from anyone.

In addition to ABO blood types there is another marker. Rh positive and negative. Anyone working with pregnant women should know about this.

If the mother is Rh negative and the father Rh positive, the fetus could be Rh positive. During delivery or shortly before, some of the baby's blood would enter the mother. If the baby is Rh positive her body would produce antibodies against its blood cells.

This does not usually affect the first delivery. Since the baby will be delivered before the mother produces antibodies. But if she wants a second baby her immune system would destroy the baby's blood cells and it could miscarriage.

To prevent this her doctor would inject her with an IgM antibody that would destroy any fetal blood cells within her system, within three days. This would prevent immunity and the IgM would not affect the fetus because it cannot go past the placenta.

Then there are white blood cells, which destroy foreign invaders, dead cells, and cancer. There are three basic types:

1. Those that attack directly, such as neutrophils.
2. T-helper cells that package foreign antigens, take them to the thymus in the chest, and present them to T-killer cells that circulate looking for a target.
3. Lymphocytes that form B cells that transform into plasma cells. Plasma cells contain RER that synthesize water-soluble proteins: antibodies.

Finally there are the platelets that allow blood to clot. Injured tissue releases chemical messengers that trigger the extrinsic and intrinsic coagulation systems. They cause a water-soluble protein called fibrinogen to become a water-insoluble protein called fibrin. Fibrin forms a net over the injury, and platelets adhere to it. This creates a clot that stops the bleeding.

Some males suffer from an X-linked disorder known as hemophilia in which the cascade systems are defective and clotting is impaired. The most common forms are von Willebrand factor deficiency, in which the patient does not produce clotting factor number eight or produces a defective factor. Less common is the "Christmas disease," hemophilia B, with a lack of factor number nine. Some patients also suffer from pancytopenia and do not have enough platelets. They might need a platelet transfusion and platelets typically last seven days.

The Renal System

Let us move on to the renal system.

Most people are born with two kidneys. They are two centimeters to either side of T12. They have a lima bean shape and are the size of an adult fist. The upper pole is covered by the ribs and the lower pole is not. They generate two liters of urine a day and are 10-20 times more efficient than dialysis machines.

We measure the activity of kidneys in terms of glomerular filtration rate (GFR). They normally filter 60,to100 mL of serum per minute. They are usually at the peak of their performance when the patient is twenty-five years old and lose 1 percent of their function every year thereafter. One of the earliest warnings of kidney failure is a UA that reveals microalbuminuria. It suggests that they are being damaged and need to be investigated.

At this point we should differentiate between acute and chronic renal failure. Acute failure usually occurs over a short period and is preceded by a catastrophic event:

- Shock

  Blood loss
  Infection

  Acute drug intoxication (such as Bactrim DS, ibuprofen, some anticancer drugs, and contrast dyes used in radiology)

Acute renal failure is often reversible and GFR will often improve once the offending condition has been removed.

Chronic renal failure is preceded by long-term pathology such as diabetes, high blood pressure, or an autoimmune condition. It is usually progressive and nonreversible.

The kidneys arise from mesoderm, and we cannot clone a reliable organ. The right kidney is 2 cm lower than the left, and the loop of Henle (within the nephron) is the unit of filtration.

Renal Addendum

A small number of people are born with one kidney. This is usually not a problem and they may go through life never knowing it. Some are born with a "horseshoe" kidney—a large singular kidney that extends left and right and usually does the work of two kidneys.

Some of the medical conditions that affect kidneys include:

- Diabetes

  High blood pressure

  Consumption of certain medications such as:

  > Anti-inflammatory (i.e., Motrin, naproxen)

  > Some of the antibiotics such as polymyxins or aminoglycosides

  > Some anticancer drugs such as Cis-platinum

  > Certain herbal medications

  > Occasionally, contrast dyes used in radiology

- Autoimmune conditions occasionally triggered by strep throat (and others)

A patient who has been exposed to the above should have a urinalyses and a renal blood test to assess renal function.

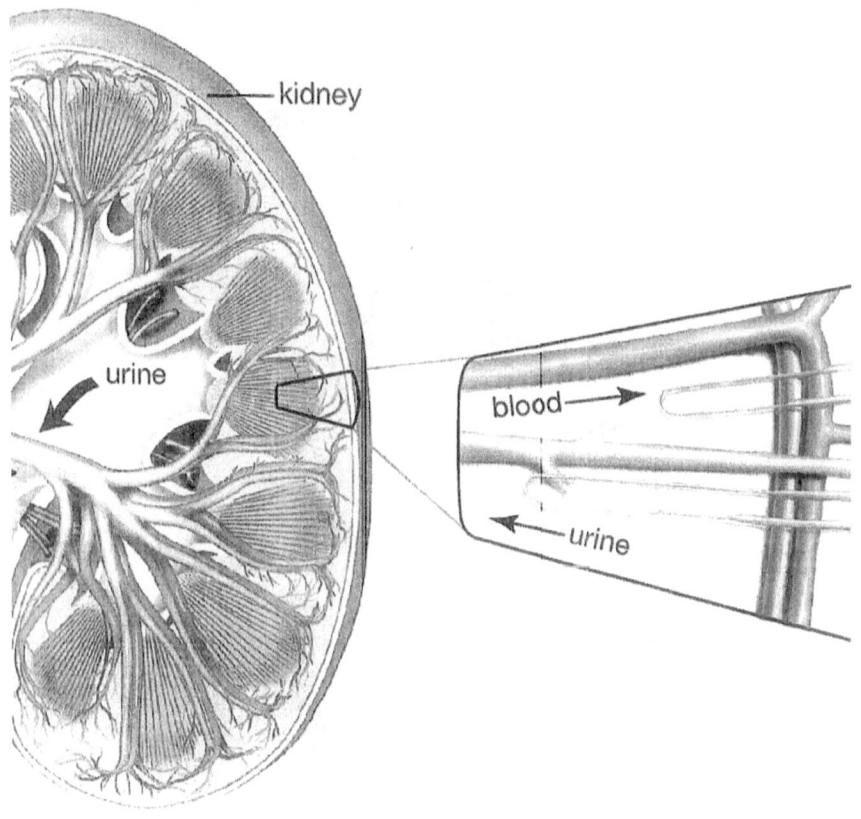

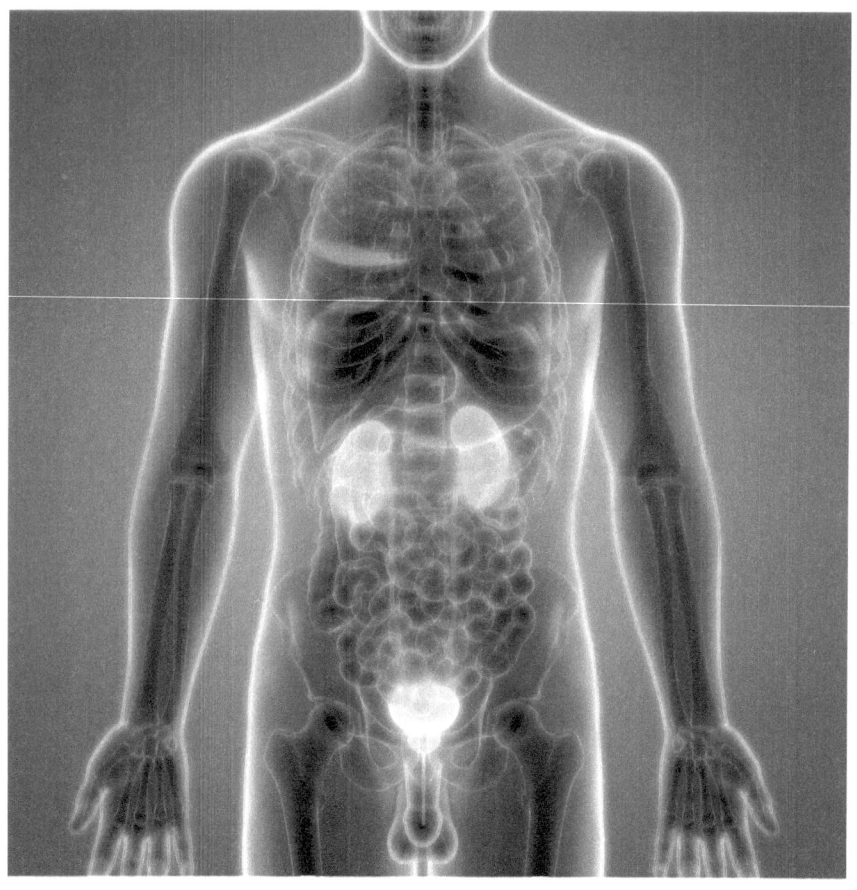

Kidneys and Ureters

Human Kidney Anatomy

External View **Internal View**

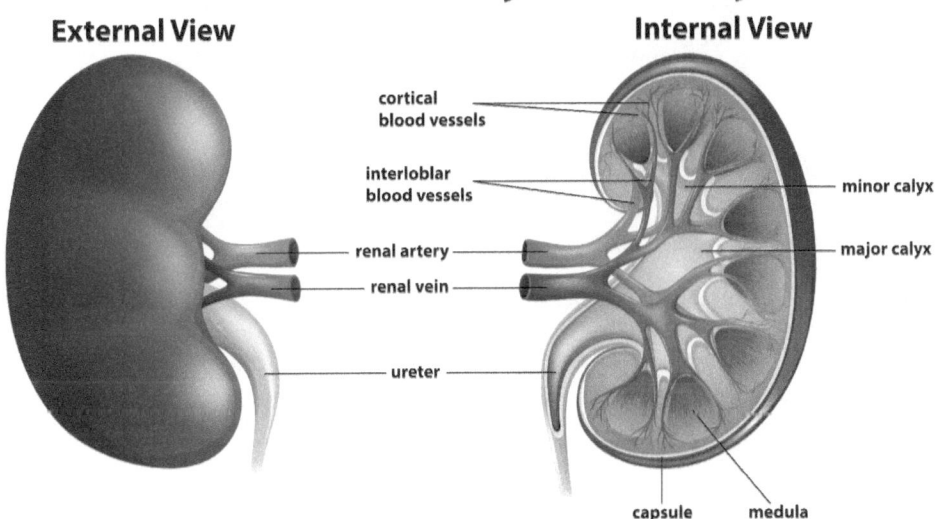

cortical blood vessels

interloblar blood vessels

minor calyx

major calyx

renal artery

renal vein

ureter

capsule medula

Kidney Anatomy

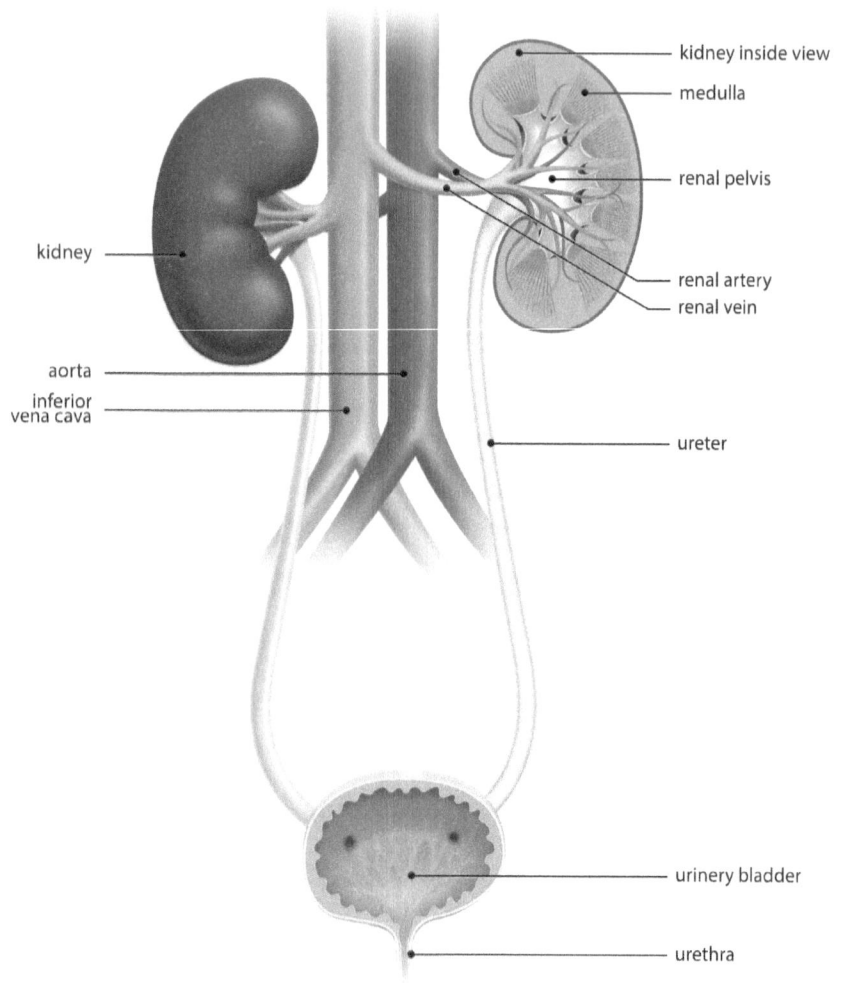

kidney inside view

medulla

renal pelvis

kidney

renal artery

renal vein

aorta

inferior
vena cava

ureter

urinery bladder

urethra

Skin

Skin is an organ, a collection of tissues working together to perform a common function. It consists of three layers:

1. Keratin: a layer of dead skin cells
2. Epidermis
3. Dermis

Keratin is a superficial layer and consists of dead cells.

Epidermis is the next layer and consists of living cells but is devoid of blood vessels or lymphatics.

Dermis is the deepest layer and contains blood vessels, lymphatics, and nerve endings.

If a melanoma develops, it may be curable as long as it is confined to the keratin or epidermis. If it reaches the dermis, it becomes incurable. It metastasizes with the blood vessels and the lymphatics.

The human body and many other higher life forms consist of organ systems that work together to make them adaptable and resilient. We are evolving into different life forms all the time.

Ecology

Let us move on to our next topic, ecology. Ecology is a study of the earth's 17 ecosystems:

- The Arctic and Antarctic (North and South Poles)

    Tundras
    Chaparrals
    Alpine forests
    Deciduous forests
    Rainforests
    Savannas
    Grasslands
    Meadows
    Ponds
    Swamps
    Marshes
    Tide pools
    Oceans
    Seas
    Deserts
    Lakes
    Salt flats

These are discrete regions where specific life forms dwell. They have adapted to their environments over thousands of years. There are some very small ecosystems such as pipes and caves, but most cover vast regions.

Let us define each:

The polar regions are covered with ice and snow beyond boundaries known as the Arctic Circle and the Antarctic Circle (near the North Pole and South Pole respectively).

The life forms of the North Pole include:

• Polar bears

  Arctic foxes
  Musk oxen
  Walruses
  Seals
  Seabirds
  Reindeer
  Whales
  Fish

All of these life forms have adapted to a cold environment. Their bodies contain copious amounts of fat insulation, and they use proteins and lipid molecules that remain liquid at low temperatures. These molecules often contain double bonds (are unsaturated), which greatly lowers their freezing point.

The polar bear can run 48 mph to catch an occasional reindeer. They can also swim sixty miles in pursuit of seals and can live for a month without eating. A polar bear typically finds food on one out of every twenty hunting trips. They will nurse their young for five weeks, but will struggle to find food for them before they starve.

The arctic fox lives in symbiosis with the polar bear. They clean up the mess that the bear makes while eating. In exchange they get a free meal and protection. Occasionally the bear will try to drive them off. But they are fast and nimble and stay out of the way. They have a thick white coat that keeps them warm and serves as camouflage.

Musk oxen have short stubby legs and thick coats that allow them to tolerate arctic weather year-round. Like all oxen they eat vegetation. Their short legs cost them speed, and they cannot run as fast as their bigger cousins (although musk oxen are actually more closely related to sheep and goats than to oxen).

Seals, sea lions, elephant seals, and walruses are members of the pinniped family. They are a favorite meal for polar bears, but most bears will stay away from walruses. Walruses weigh 2,000 pounds, have six-inch tusks, and can cripple a polar bear. Seals are the prey of choice. But as fish migrate north, so do the seals.

Penguins live at the South Pole and can swim for 5,000 miles by allowing half of their brain to sleep at a time. They can dive to 2,000 feet, and even though they cannot fly, they can jump off a cliff 300 feet high without injury.

During mating season females will swallow fish, walk seventy miles, and feed the babies and the males.

Penguins and seals also have a kind of hemoglobin (called myoglobin) that holds greater amounts of oxygen for a longer period while they make deep dives. Life at the North and South Poles is a struggle until the aurora appears. It indicates that summer is coming.

Seabirds eat mostly fish.

As is true of most antelope, reindeer can jump across a ravine fifteen feet wide and over a fence fifteen feet tall. In 1823 Clemente C. Moore thought that they could fly and wrote a poem about eight enchanted reindeer who pulled Santa's sleigh at Christmas.

The food supply at the North and South Poles depends largely upon fish. They in turn need water that has oxygen. As the earth's oceans begin to warm, fish have to migrate farther and farther toward the north and south to find oxygen. This is because gases (including oxygen) become less soluble in water as the temperature rises.

The future of the earth's arctic and antarctica regions is in jeopardy. They continue to warm, and the ice continues to melt.

The next ecosystem is the tundra: vast cold plains that are covered with short stubby herbs and shrubs. It is home to a wider variety of plants and animals, birds, deer, and bears.

One of the more interesting animals is the elk a large deerlike creature that weighs 500 pounds. They can run 60 mph over short distances and maintain 30 mph over longer distances. The reason for this is that prior to the earth's last ice age, they had to escape cheetahs. And even though the cheetahs are long gone, their ability to run is not. This does, however, allow them to escape another predator that shares the tundra, the Kodiak bear. Close relatives of the polar bear, they have been clocked at 50 mph.

Another very fast-moving animal from this region is the jackrabbit, which has been known to run at 45 mph. So speed protects many from the ravages of predators.

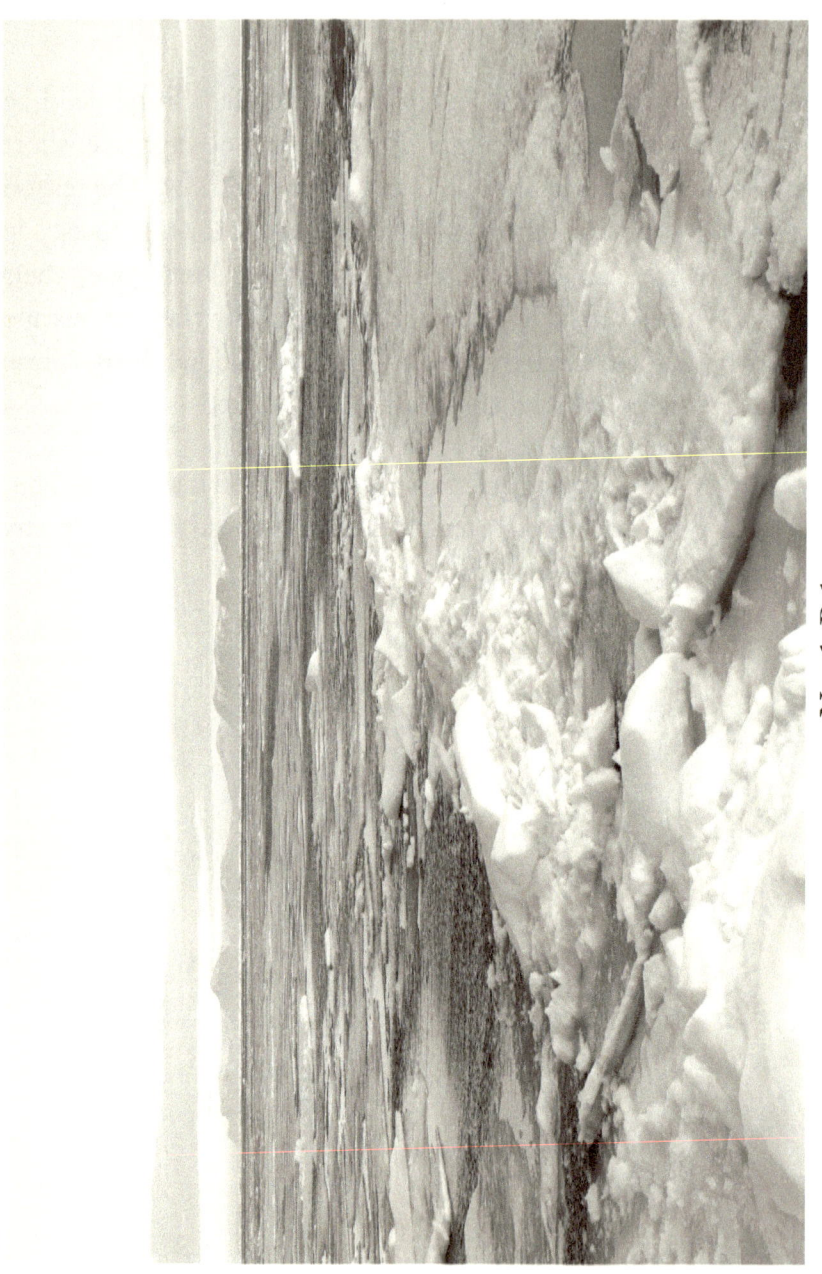

North Pole

Polar bear

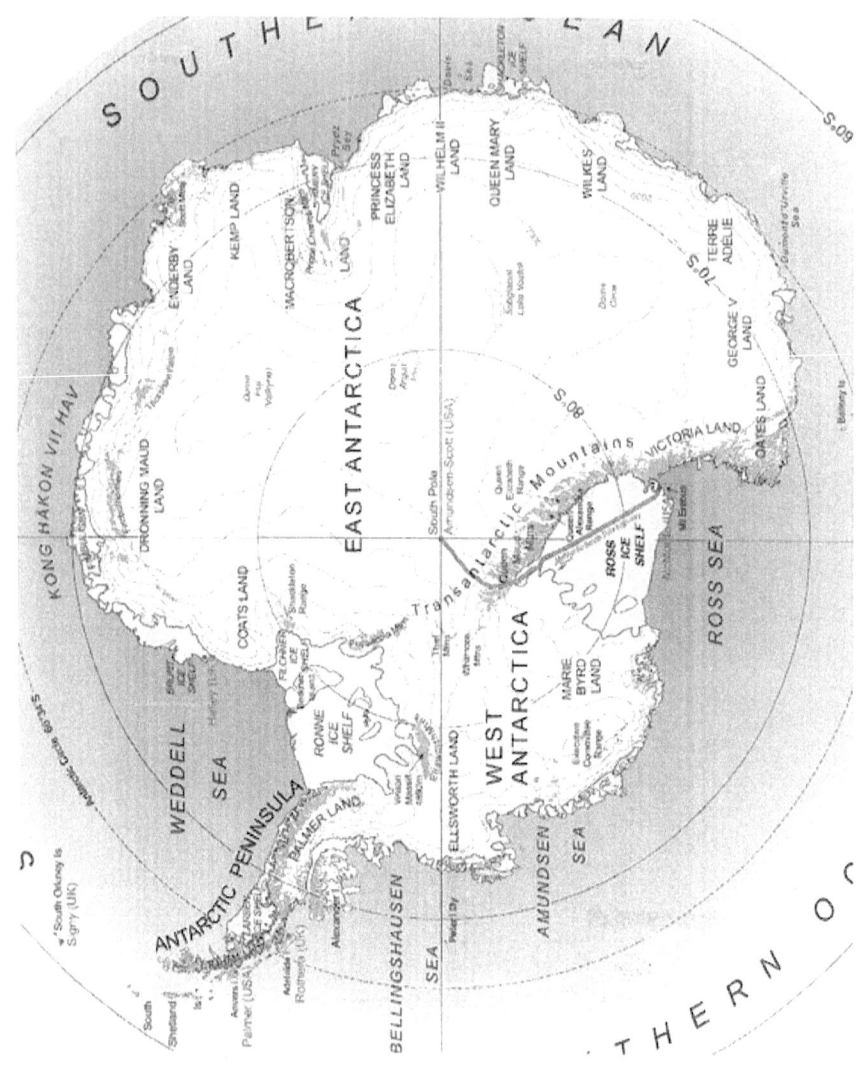

Kodiak Bear

The vegetation of the tundra grows slowly. If a jeep drove over it, the imprints would remain for more than fifteen years, so they must be protected by strict laws.

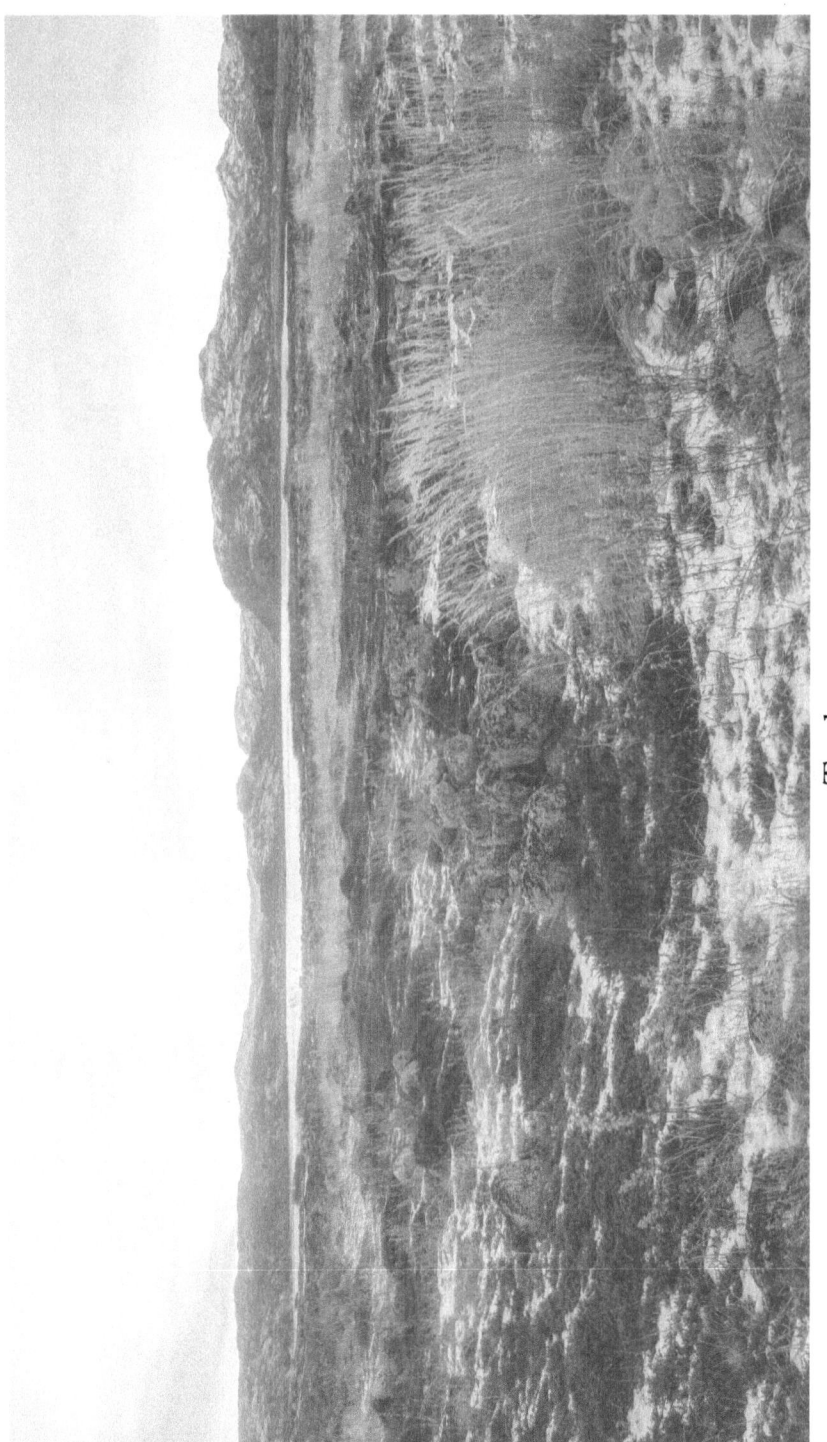

Tundra

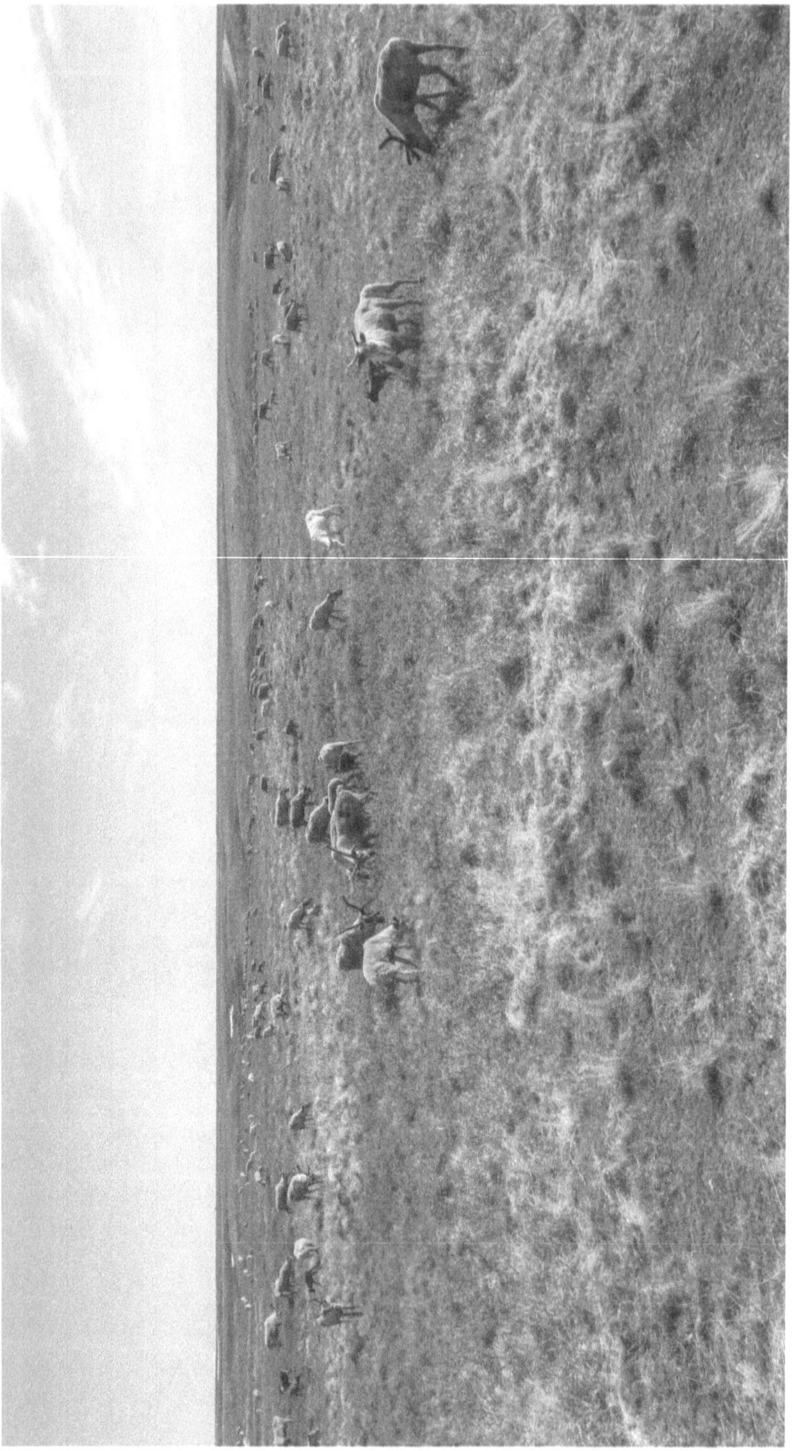

Tundra

Chaparral

Chaparrals are characterized by dense vegetation consisting of bushes and low-lying trees. They typically occur at altitudes of 5,000 feet or higher.

Chaparrals are home to foxes, a variety of cats, birds, rabbits, and snakes, and many plants including manzanita (which means "small apple" in Spanish). When the Spanish conquistadors of the 1600s first saw them, they thought they were small apple trees.

Although the terrain is rugged, chaparrals can be used to raise cattle and other domestic animals. In order for them to perpetuate, they need to be set on fire once every 10–12 years. This serves two functions.

- Thins out the underbrush

  Generates the heat needed to make some plants release their seeds

Most chaparrals also grow on the sides of mountains.

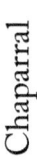

Chaparral

The Chaparral Biome

As is true of most regions with predator and prey, the predators only have a slight advantage over the prey. This allows the balance to be maintained in which all animals get enough to eat. As predators get older, they may be forced to prey on slower-moving animals or to eat carrion (scavengers). But what is amazing about the tundra is that many of these animals have retained the characteristics they once needed during the last ice age.

Alpine Forests

Alpine forests typically thrive in cold weather. Some of the trees lose their leaves during the winter months. Others do not and are referred to as evergreens. Evergreens are often used as Christmas trees and are home to many life forms, including birds and chipmunks. They also provide a constant source of food.

Alpine trees typically grow needles and cones. Needles are modified leaves that help to conserve water, and cones store seeds. Alpine forests are home to:

- Many rodents including chipmunks

   Many species of birds including the red cardinal

   Some species of deer

   Some members of the weasel family, including wolverines, badgers, and ferrets, all of whom are extremely aggressive.

- Raccoons

- Bears (relatives of raccoons)

  Several species of cats such as the bobcat
  Foxes
  Wolves

Next are the deciduous forests. Deciduous forests receive between 75 and 100 inches of rain per year. There are many on the North American continent.

Evergreen Forest

Deciduous Forest

Rainforest

Addendum

Let us talk further about deserts. They are regions that are close to the earth's equator and receive less than ten inches of rain per year. They are as hot as jungles, but most are surrounded by mountains that block the passage of clouds. As a result, they receive little rain.

The life forms have adapted to high temperatures and low availability of water. They include:

> Cacti: their leaves come in the form of needles that allow them to conserve water. They have thick integuments (skins) that are surrounded by a wax called cerumen. They have a barrel-shaped body habitus that is ideal for conserving water, and they have a mat of roots near the surface that is efficient at collecting water during occasional rainstorms.

> Snakes: tend to be motionless during the day and usually underground. At night they use heat pits under their eyes to see infrared (heat). This gives them night vision (the ability to see in the dark).

> Cats are active at night. Their eyesight is six times better than that of a human and they have reflective material inside their eye orbits which makes it easier to see in the dark.

> Insects and scorpions bury themselves in the sand by day and become active at night.

> Animals such as the camel can live in a desert for almost a month without drinking. They generate water inside their bodies through different chemical reactions.

Some birds such as condors use hot-air thermals that rise from deserts to soar at high altitudes, where temperatures are lower for hours. They turn their bodies into gliders by using postural muscles. These muscles are dark red and contain a high concentration of mitochondria, which provide energy in the form of ATP.

There are only a handful of ecosystems left.

- Marshes

  Tide pools
  Salt flats
  Seas
  Oceans

Marshes are sometimes referred to as wetlands. They are the transition between freshwater rivers and the ocean. To a chemist freshwater is anything that is (1.75) percent sodium chloride by weight or less. This is the only criterion and this is what characterizes marshes. They are 1.75 percent salt.

They are home to:

- Birds

  Fish
  Rodents
  Wide variety of plants

Tide pools are the transition zones between beaches and oceans. They:

Are rocky

Contain seaweed and sea grass

Have a higher oxygen content than ocean water

Are homes of many crustaceans such as hermit crabs, crabs, lobster, shellfish, and periwinkles

Their sharp surfaces provide protection against sharks.

They are inhabited by life forms that thrive in waters that are in constant motion.

Salt flats, such as the flats of Utah, are arid and largely devoid of life.

The word *sea* is not used much anymore. It is an antiquated term that refers to a large body of salt water such as the Adriatic Sea and the Sea of China. But most of the life forms are similar to ones found in the oceans, and most naturalists only speak of the Atlantic Ocean, the Pacific Ocean, the Indian Ocean, and the Arctic Ocean.

We can characterize oceans as:

- Covering 3/5 of the earth

  Providing the earth with oxygen and food

  Maintaining a consistent temperature

  Maintaining a constant supply of water

  Providing the human race with a safe means of eliminating sewage

  The source of some violent storms such as hurricanes

  Once provided accessible routes to most parts of the world

The deepest point in the ocean is the Challenger Deep in the western region of the Pacific. It is 36,200 feet (or approximately 7.5 miles) deep.

It also contains underwater mountains, valleys, trenches, and volcanoes.

The oceans are home to a wide gamut of life all the way from single-celled organisms to the largest creatures that ever lived, including the dinosaurs and the blue whale, weighing in at 190 tons.

Whales once lived on land and looked like giant seals. Then they migrated back to the ocean. They are close relatives of two very intelligent animals: dolphins and porpoises. Like their smaller cousins, they have sonar and a blowhole. Whales can swim thousands of miles. They do this by allowing half of their brain to sleep at a time.

Ocean water has four characteristics that are worth mentioning:

1.
2. It contains small amounts of every element in the world (all 106 of the elements on the periodic table).
3. 2.
4.
5. For every ten feet that you go down, it absorbs 90 percent of the sun's light. So at ten feet only 10 percent remains, and at 20 twenty feet only 1 percent, and so on.

One could calculate the intensity at a given depth with the following equation:

$$iI \text{ (new)} = i \text{ (surface)} * e < \text{neg. } 0.693 \text{ (depth/HVL)}>$$

Intensity (at depth) = intensity (at surface) x e raised to the negative 0.693 x depth/half value layer, when half value layer equals 3.92

feet. After passing through 3.92 feet of ocean water, half of the sun's light is gone.

1. 3.
2. It is extremely effective at killing dangerous bacteria and at breaking down most forms of human refuse (with the exception of plastic).
4. .The salinity (concentration of salt) stays constant at all temperatures. So the water at the South Pole has the same salinity as water at the equator, and the concentration of the salt does not change regardless of the temperature. This is because sodium chloride is a euthermic salt. If you dissolve it in water, the temperature will not change.

Marsh lands

Tide Pool

Great Blue Whale

Ocean

Jacques Cousteau

"The oceans are warming up and the polar
ice caps are melting. We need to act."

As the temperature of water changes, the concentration of salt remains stable. To understand this we need to discuss new concepts. All water-soluble salts have a crystal lattice energy and an energy of hydration.

Lattice energy is the energy needed to break up the salt's crystalline structure. When it dissolves it will form anions and cations, which will combine with water molecules. When they do, they produce energy which is the energy of hydration.

If crystal lattice energy is greater than the energy of hydration, it will be an endothermic salt and will cool the water.

If crystal lattice energy equaels the energy of hydration, it will be a euthermic salt and will neither cool nor heat the water.

If crystal lattice energy is less than the energy of hydration, it will be an exothermic salt and will heat the water.

i.e. NaCl $\xrightarrow{H_2O}$ Na+ + Cl- + H2O $\xrightarrow{\quad}$

In chemistry heat can act as a reactant or as a product and is represented by the letter Q.

Under le Chatelier's law, if we add a reactant or remove a product, the reaction will run faster and with greater ease.

Let us say that Q + A + B $\longrightarrow$ C: it would run faster in hot water because we add Q as a reactant.

Endothermic salts dissolve with greater ease if we mix them with hot water.

Let us say that A + B —→ C + Q. An exothermic salt will dissolve with greater ease in cold water because we take away the product Q.

Let us say that A + B -→ C. A euthermic salt would dissolve in water of any temperature with the same degree of ease because we neither add a reactant nor remove a product. This is true of NaCl or table salt.

So if a salt is endothermic (makes water cold), it will dissolve more rapidly in hot water where we supply heat as a reactant:

Q + A + B-> C + D runs faster because we added heat to the reaction.

And exothermic salts (which make water hot) will react more rapidly in cold water because the water takes away one of the products, heat.

X + Y -> Z + Q

But if a salt is euthermic (neither cools nor warms the water), then the amount that will dissolve does not change appreciably as the temperature changes.

So if a salt NaCl -> Na+ Cl- with no Q, then ambient temperature will have little effect upon its solubility or its concentration. So ocean water at the South Pole and at the equator will have virtually the same salinity, and surface water is going to have almost the same salinity as water at great depths.

This is particularly relevant to whales, dolphins, penguins, sharks, sea stars, and schools of fish who dive very deep and swim great distances (thousands of miles). It would not be possible if salt concentrations changed appreciably.

Another aspect about the chemistry of ocean water is that it is a good buffer: it resists changes in pH. This is particularly important to shellfish, who find it progressively difficult to create a shell as pH drops (or becomes more acidic). Shells are solidified calcium carbonate and will not solidify in an acid environment. So as pH drops, it becomes more difficult for them to form a shell and in recent years this has been a problem. The concentration of gases in the atmosphere is causing bodies of water to become more acidic. These gases are acid anhydrides, and carbon dioxide is one of the biggest offenders. If pH gets too low, it affects the enzymes of living things.

One of the most interesting life forms in the ocean are the kelp beds. Like extensive underwater forests, they are the homes and source of food for many marine animals. As we said, the oceans are effective at killing harmful bacteria and at providing oxygen. Oxygen is one of the key players in allowing higher life forms to develop. The more efficient aerobic respiration provides the energy needed to create and maintain multicelled organisms.

Microbiology

Let us move on to another form of biology: microbiology, the biology of organisms too small to see with the naked eye.

There are four names that deserve recognition:

Antonie van Leeuwenhoek, Percivall Pott, Joseph Lister, and Louis Pasteur.

Leeuwenhoek was a Dutch cloth merchant who had a passion for grinding lenses, especially convex lenses used in eye glasses. In 1676 he built the first microscope using one lens. It was like a magnifying glass and had a magnification factor of 40.

He made careful drawings of microscopic life forms living in a drop of pond water and became known as the father of microbiology.

Antonie van Leeuwenhoek

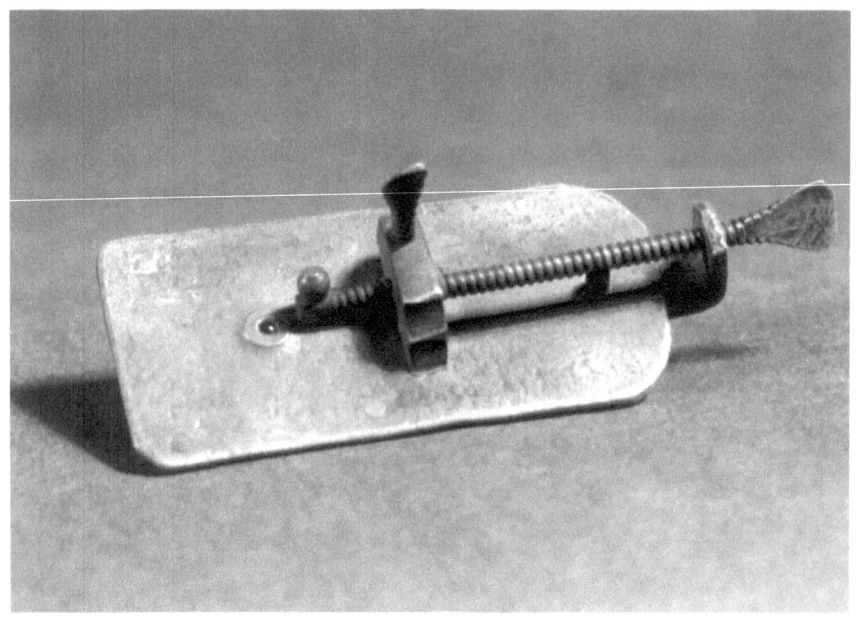

Leeuwenhoek
Microscope
(circa late 1600s)

Leeuwenhoek's original microscope used one lens. It was more like a magnifying glass than a modern microscope.

Modern microscopes have two convex lenses. They also have a focusing tube. They focus light onto a focal point and then onto the second lens.

Let:

O = size of object

L = length of focusing tube

F = focal length

I = size of image

$$\frac{O}{F} = \frac{I}{L}$$

Therefore:

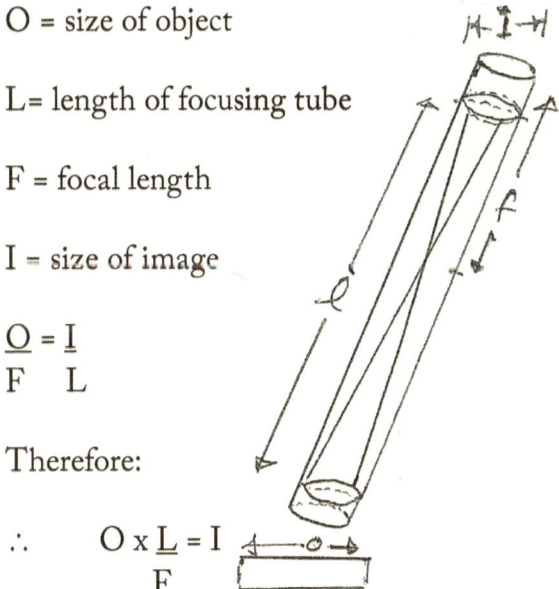

$$\therefore \quad \frac{O \times L}{F} = I$$

Object size (O) x length (L) divided by focal length (F) equals size of image (I).

So if the tube is 6 cm (60 mm) and the focal length is 1 mm the magnification factor would be 60/1 = 60.

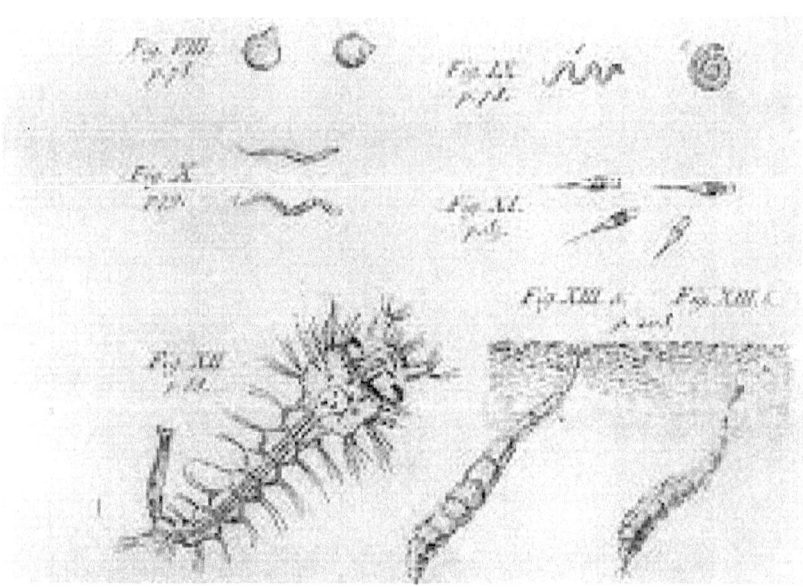

Antonie van Leeuwenhoek studies microscopic life forms in a drop of pond water:

Percivall Pott, F.R.S.

Percivall Pott was a physician in the eighteenth century who treated smallpox. It was a deadly disease that killed many. But he noticed that cow maidens (who milked cows) rarely contracted it. He discovered that they caught cowpox (a mild variant of smallpox) and became immune. He removed scabs from cows that had it and placed them into a small incision made in humans. It successfully inoculated them and saved millions of lives.

Louis Pasteur

Louis Pasteur was a chemist who lived in the nineteenth century. He created vaccines, sulfa-based antibiotics, and the process of pasteurization, which saved countless lives.

The Chinese inoculated people thousands of years ago. They removed scabs from patients who had recovered and placed them into the noses of people who needed to be vaccinated.

Joseph Lister

Joseph Lister was a British surgeon who lived in the late nineteenth century. He convinced his fellow surgeons that they had to scrub and gown before operating. He saved countless patients from sepsis and death.

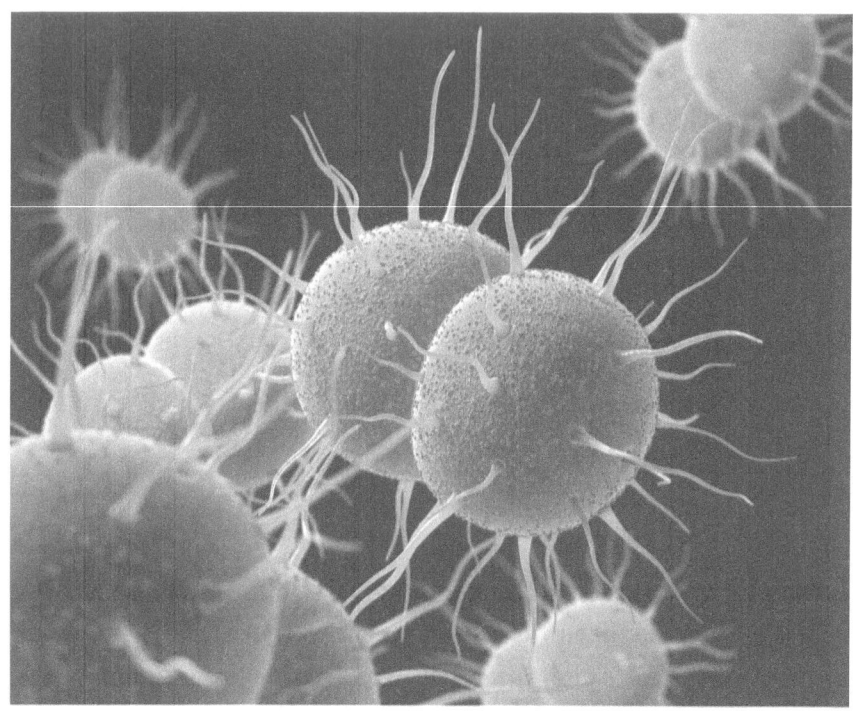

Diplococcus

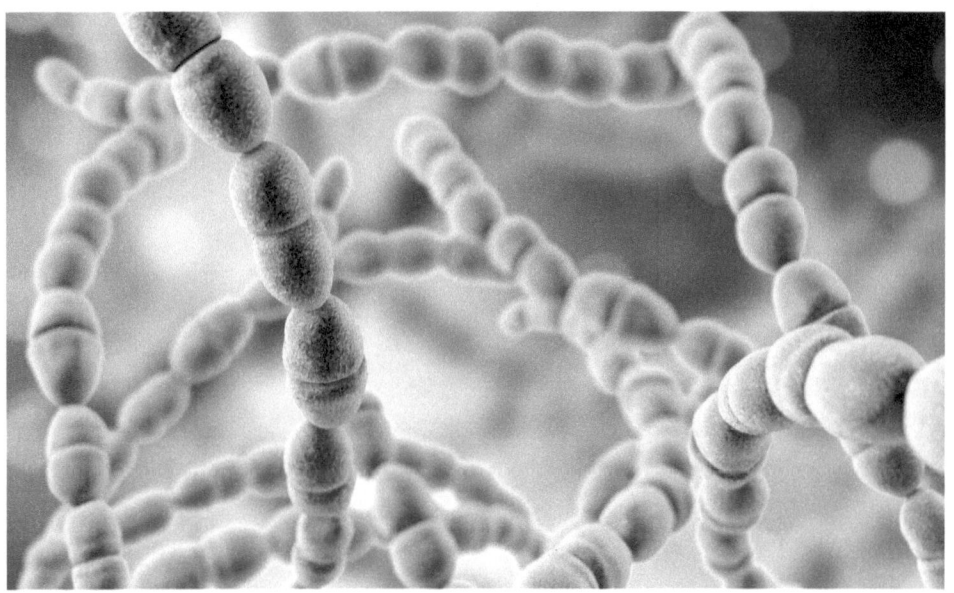

Streptococcus

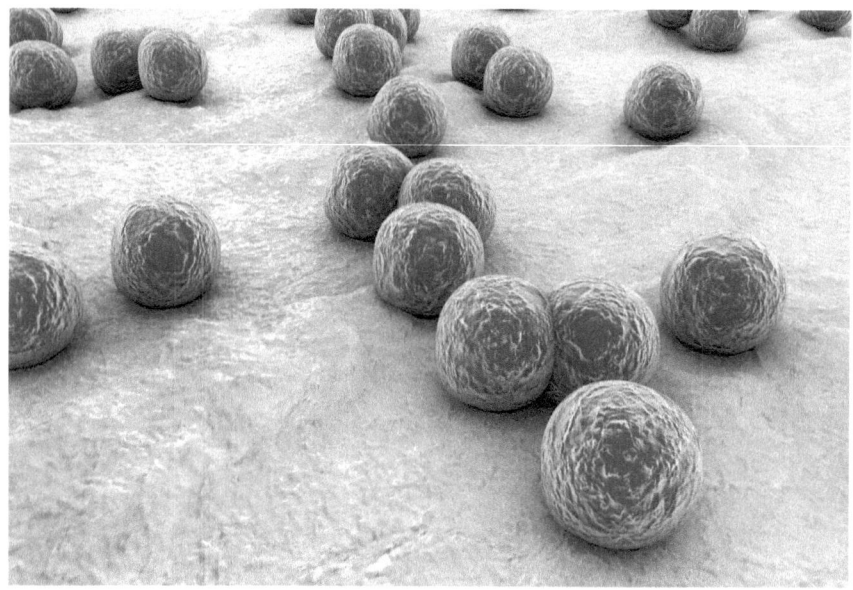

Staphylococcus

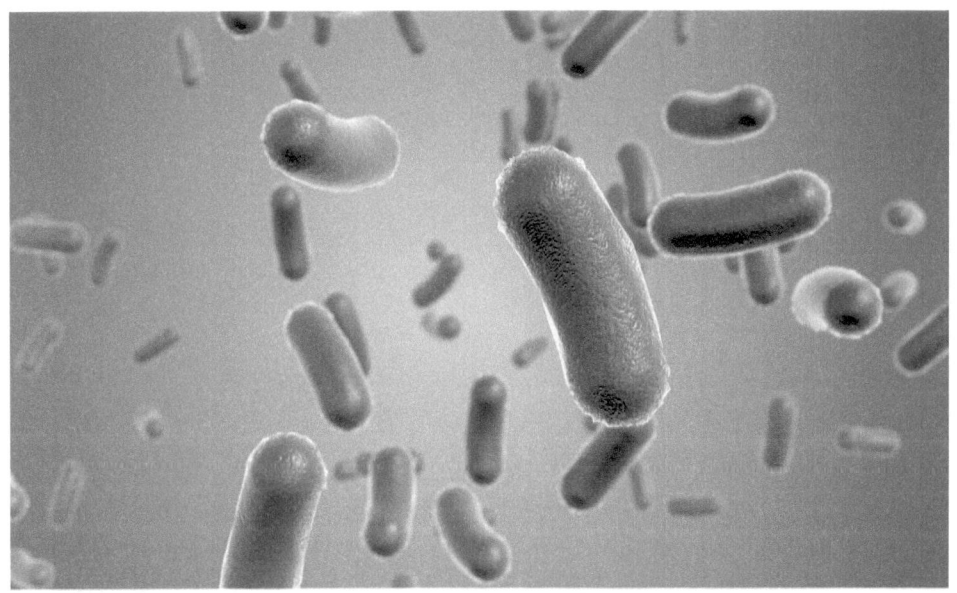

Streptobacilli

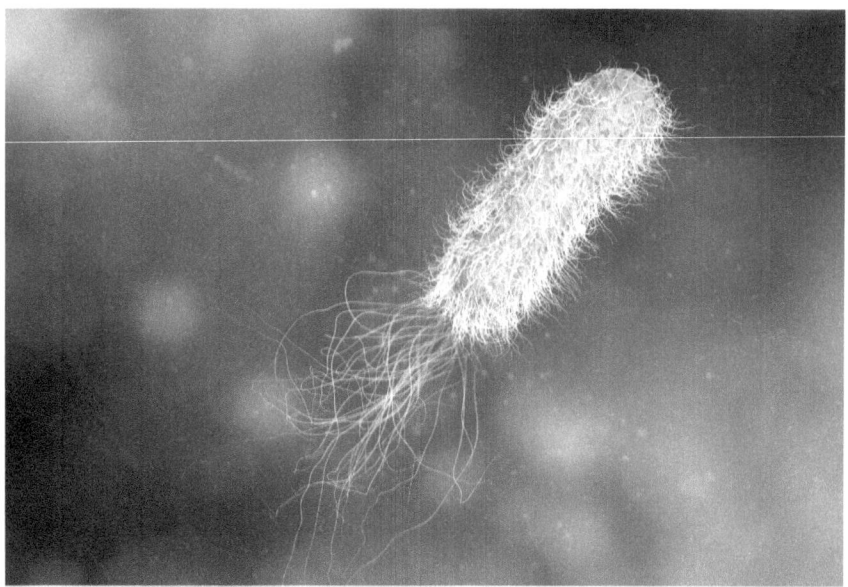

Flagella

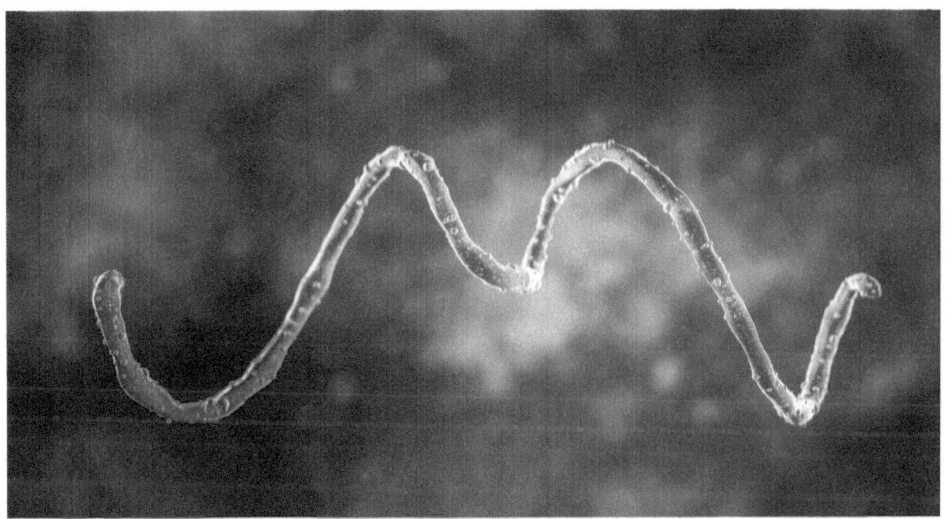

Spirochete bacteria

Addendum

Let us look at some of these organisms and discuss specific diseases that they cause.

Diplococcus causes gonorrhea. Diplococcus means "two spheres" in Latin.

Streptococcus can cause strep throat. Streptoccus means "a string of spheres" in Latin.

Staphylococcus can cause wound infections. Staphylococcus means "bunch of spheres" in Latin.

Bacilli can cause tuberculosis and typhoid. Bacilli means "sticks" in Latin.

Flagellated organisms can cause cholera. Flagella means "with a tail" in Latin.

Spirochetes can cause syphilis. Spirochete means "like a coil" in Latin.

Types of Viruses

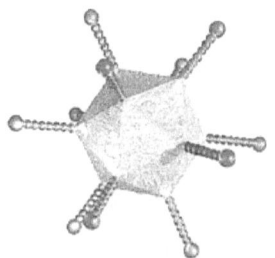

Adenovirus

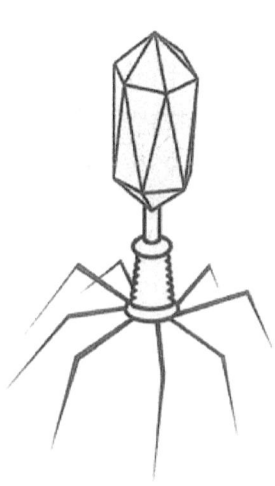

Bacteriophage   Human Immunodeficiency Virus

When viruses infect a cell they do two things:

They insert their DNA into the cell and take over its DNA and protein-synthesizing machinery. Some viruses insert viral RNA and use an enzyme called RNA reverse transcriptase to make viral DNA. The cell then makes a viral DNA or RNA and the virus's protein coat.

There are two kinds of viruses: viruses that leave the cell through reverse phagocytosis and viruses that cause a cell to burst open and release newly synthesized units. As you might imagine, the blastic viruses are far more damaging. This accounts, in part, for the fact that some viruses are more virulent than others.

All viruses are smaller than the wavelength of light making it impossible to see them with the light microscope. If you want to know if someone has a virus there are three things you can do:

You could test their blood serum and see if they're making antibodies against the virus. You take a liquid that contains the viral antigens (proteins that appear on the surface of the virus) and mix it with the patient's serum. If antibodies are present, you get an antibody-antigen reaction. This would be a Lewis acid, and if you mix it with a litmus indicator, it turns red.

Another way is to take a liquid that you think contains viruses and mix it with cells suspended in a liquid matrix. You can look at the cells with a microscope and see whether or not they are undergoing changes indicative of viral infection. In the example, do you see phagocytosis or blast activity?

The third way is to visualize them directly using an electron microscope. These devices use beams of electrons instead of light. You can see much smaller objects because the wavelength is much shorter. Electrons form de Broglie waves as they travel through space. They have an extremely small wavelength.

Viruses and some bacteria also go through cycles. Chickenpox (varicella) is a classic example.

When a patient is first infected, there's a pulmonary phase, in which they will develop a dry cough. This typically lasts one or two days and is followed by a hematogenous phase, in which the virus enters the blood and causes fever and malaise (a general sense of weakness and not feeling well), which lasts another two to three days.

Then there is a cutaneous phase in which the patient develops pruritic (itchy) pustules over most of their body.

Finally there is a neurological phase. The virus enters nerve fibers, which course with the skin dermatomes and follow them back to the spinal ganglia. There they stay until they are reactivated by:

1. Exposure to sunlight
2. Cold weather
3. Emotional stress
4. Anything that suppresses immunity (such as treatment of lupus, rheumatoid arthritis, psoriasis, cancer, or an organ transplant)

Then they leave the ganglia, reenter the nerve fiber, and reinfect the skin usually on the chest wall. The skin will become tender and will break out in shingles (a crop of painful lesions). Shingles is a form of herpes and can be treated with antivirals:

1. Acyclovir
2. Valtrex
3. Famvir

Long-term control can be achieved with two doses of herpes vaccine. It will reduce the frequency, severity, and duration of future attacks but will not completely stop them. Occasionally herpes will damage local nerve fibers and cause postherpetic neuralgia. This leads to severe lingering pain and can be treated with:

1. Lidocaine patches
2. Tricycles

Varicella is one of twenty-one herpetic viruses.

Let us change direction and talk about fungal infections:

1. Coccidioides
2. Valley fever
3. Pneumocystis carinii
4. Tinea pedis (Athlete's foot)
5. Tinea cruris (affects genital region)
6. Tinea corporis (skin fungus that affects skin not associated with feet, face, or genitalia)
7. Tinea versicolor (skin of face)
8. Oral candida (causes thrush)
9. Vaginal yeast

Fungi reproduce by creating spores, which are hard to kill. They can be treated with:

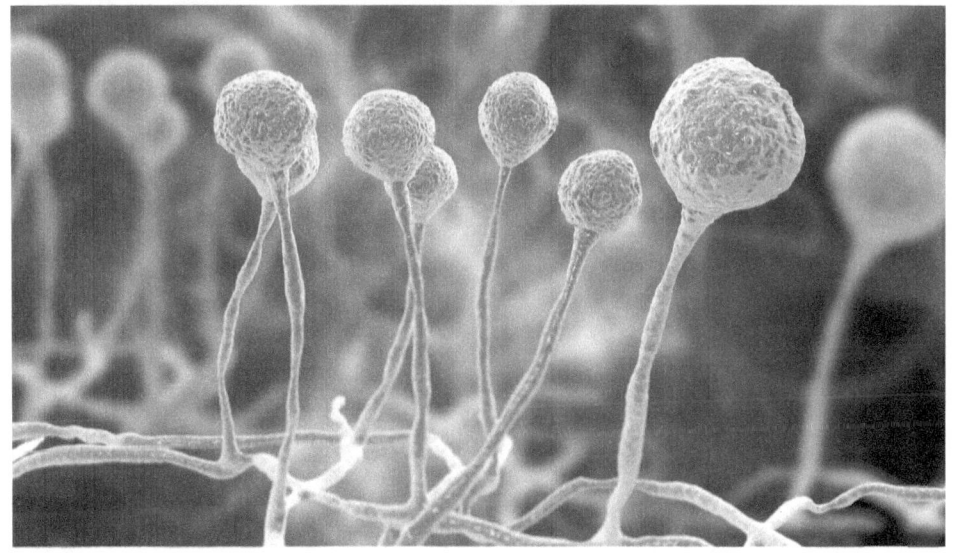

Alexander Fleming discovered penicillin

Ketoconazole
Clotrimazole
Diflucan
Lamasil
Itraconazole

How do we treat bacterial infections?

There are many antibiotics. We first need to perform a culture and sensitivity test. We get the bacteria to grow on a plate lined with blood agar. Then take paper wafers that contained different antibiotics, place them onto the plate, and look for a "kill zone" around each wafer: dead bacteria.

Antibiotics include:

- Penicillins

  Cephalosporins
  Polymixens
  Tetracyclines
  Erythromycins
  Fluoroquinolones
  Sulfa antibiotics
  Aminoglycosides

In the second phase we take the antibiotics that the organism is sensitive to and dilute them. Antibiotics that are still effective after being diluted 32 times are considered favorable.

Let us move on to another infectious organism: prions. They are protein polymers that affect the CNS. They are usually seen in third-world countries and occur in people who eat unusual animals such as monkeys.

Cooking does not stop them, and they present as:

1. Loss of cognitive function (memory and reason)
2. Profound weakness
3. Ataxia (in coordination)

They include:

1. Kuru
2. Creutzfeldt–Jakob disease
3. Mad cow disease (BSE)

There is no cure. Their presence could be verified with a CT or an MRI of the brain, and a brain biopsy. It is hard to know if prions are true life forms. They do not meet many of the seven characteristics discussed earlier. But they do behave as infectious diseases rather than as toxins. No matter how many times they are diluted, they still have a harmful effect.

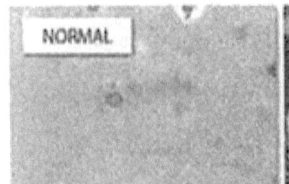
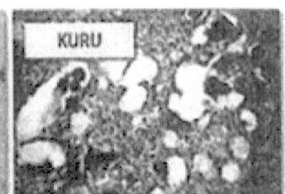

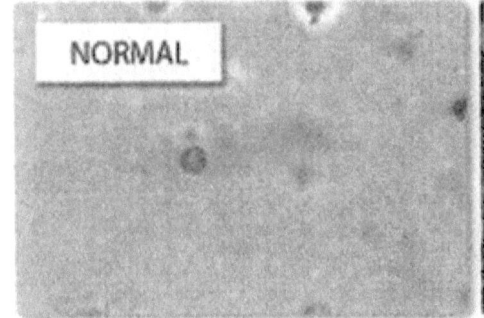
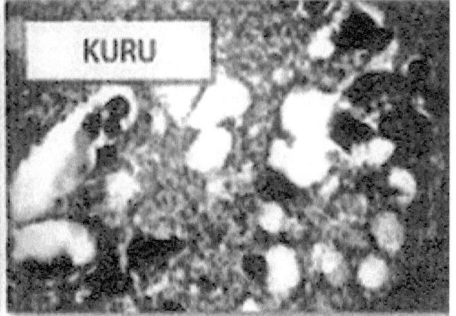

Kuru

It is hard to determine whether viruses and prions are true life forms. They do not meet many of the criteria listed earlier. They do however, qualify as infectious diseases as opposed to toxins.

To determine whether a substance is a poison or an infection, you would dilute it. You might have to dilute it thousands of times. If it is a poison, the effects would eventually wear off. But if it is an infection, it will continue to reproduce and to be toxic.

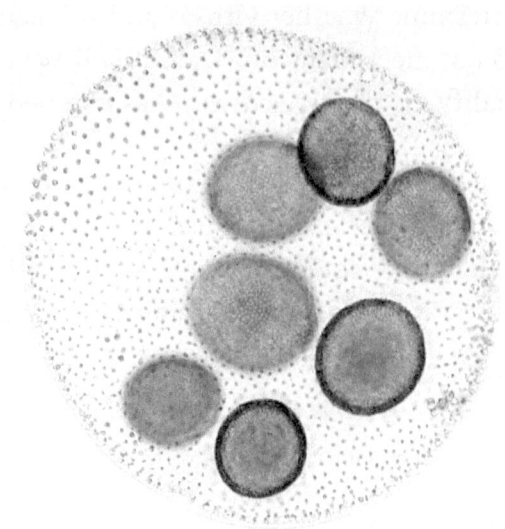

Volvox

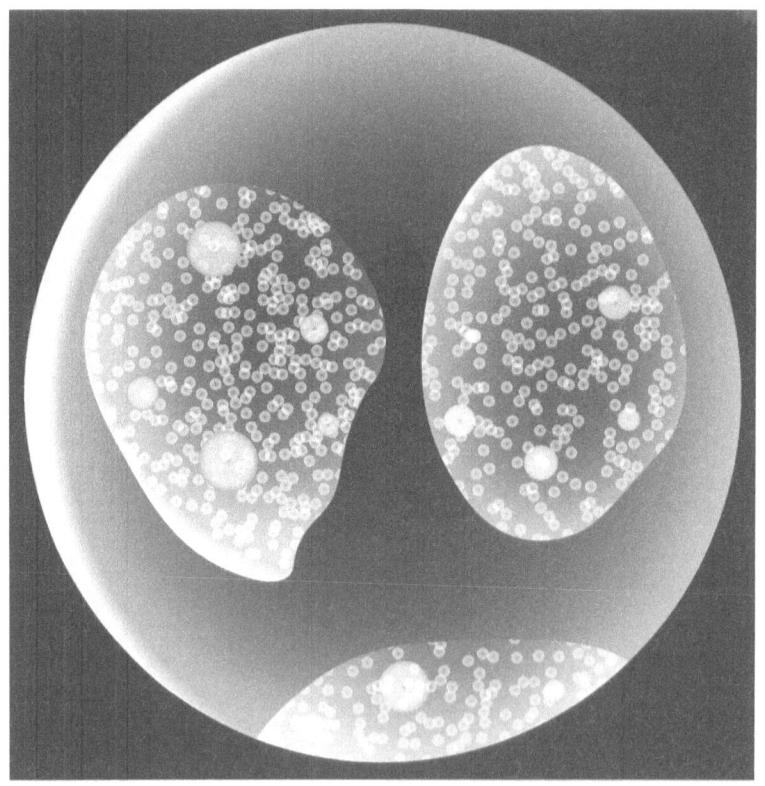

But the other organisms on the list are definitely alive. Let's characterize them one by one.

Amoebas are single-celled organisms that are mobile, cause disease, and respond to stimulus. They are usually found in water.

Protozoa are single-celled organisms that cause gastroenteritis.

Amoebas and protozoa can be treated with Flagyl 500 mg by mouth three times daily for 10 days.

Algae are single-celled organisms that carry out photosynthesis, producing food and oxygen from sunlight, water, and carbon dioxide. Blue-green algae in the oceans generate 50-80 percent of the oxygen we breathe.

Fungi break down preexisting food. Fungi can live with algae in a symbiotic relationship as lichens, which grow on rocks and trees. Fungi extract water from the air. The algae uses the water along with air and sunlight to synthesize sugar. Both organisms consume the sugar, and both extract trace minerals from the rock.

Bacteria may or may not have a cell wall. If they do, they usually adhere to Gram stain and appear as gentian violet under a microscope. If not, they will adhere to saffron and appear to be red (Gram positive and Gram negative respectively). If it adheres to Gram stain, it is Gram positive. If it adheres to saffron, it is Gram negative.

Some bacteria break down decaying matter and are referred to as saprophytes (which means "feeding on the dead" in Latin).

Volvox is a group of organisms that live together and communicate through a thread. One unit may consist of 17,000 organisms.

In medicine the most salient organisms include:

- Bacteria

  Viruses
  Prions
  Fungi

Bacteria can cause life-threatening infections such as:

- Tuberculosis

  Leprosy
  Bubonic plague
  Pneumonia
  Typhoid
  Dysentery
  Cholera
  Meningitis
  Malaria

Life-threatening viruses include:

- Rabies

  Polio
  Yellow fever
  Smallpox
  Hepatitis
  HIV

Yersinia pestis (plague bacteria) comes from rodents like ground squirrels and rats, and initially infects the lungs. The patient begins to cough and sprays droplets that contain the bacteria and may infect anyone who inhales them. Within hours it passes to the blood and causes fever, malaise, diffuse joint pain, and rapid loss of cognitive function (ability to reason and remember). If the patient is bitten by fleas, the fleas may act as vectors and pass the infection to others. Death usually occurs in two days without treatment. It can be treated with Tetracyclines.

It is rarely an issue today, but people should avoid contact with rodents, and if suspected, it must be treated quickly.

Malaria is caused by a protozoan called *Plasmodium falciparum* that usually affects people in third-world countries. It is transmitted by mosquitoes and has both a hematogenous and a hepatic phase. During the initial infection it causes malaise and high fevers. Then it becomes dormant in the liver, where it will stay for an indefinite period. Sometimes it will become active again and sometimes not, but anyone who has had it can never donate blood. The risk of contamination is too high. It can be treated (and prophylaxed) with mefloquine or chloroquine.

It is difficult to fight viruses once acquired because we don't have the equivalent of antibiotics. There are only a few antivirals:

- Acyclovir

    Famvir
    Valtrex
    Tamiflu
    Anti-AIDs drugs
    Medications for the eradication of hepatitis C

There are far fewer medications for viruses than there are for bacteria.

The best approach is to avoid contact with viruses, and if one intends to travel to a third-world country, the CDC in Atlanta, Georgia, recommends the following inoculations or vaccinations:

- MMR (mumps, measles, and rubella)

  Varicella (chickenpox)
  Hepatitis B
  Hib (polio)
  Tdap (tetanus, diphtheria, and pertussis or whooping cough)
  Influenza (flu)
  Pneumococcal pneumonia
  Prophylaxis with mefloquine or chloroquine

Most people who have been to school in the United States will have received these vaccines.

Let us take a minute to talk about what it takes to create new medications and vaccines. One has to go through three clinical trials:

In trial 1, it has to be shown that the medication can be used to treat the condition in question.

In trial 2, one has to demonstrate that it is superior to other products already on the market.

In trial 3, one has to show that it has few or no side effects, including long-range side effects occurring 10–20 years later.

The last step is time-consuming and expensive. It typically costs $1 billion to develop one medication and requires seventeen years.

Antibiotics have a variety of mechanisms. Two are common.

- Prevent bacteria from synthesizing a cell wall, which is what penicillin and cephalosporin do

Prevent their DNA from replicating, as with the erythromycins, tetracyclines, and the fluoroquinolones

Antiviral medications also have two mechanisms

- Protease inhibitors prevent cells from synthesizing a viral coat

Nuclease inhibitors prevent the cell from synthesizing viral DNA

Most of these medications come from other life forms. Molds synthesize antibiotics, and bacteria synthesize antivirals.

Antifungals are typically azols or azol-like medications that prevent fungal organisms from synthesizing a cell wall. They include:

- Ketoconazole

  Clotrimazole
  Lamisil
  Sporanox
  Griseofulvin

All of these medications work on the principle of competitive inhibition.

There are several ways for viruses, bacteria, and fungi to resist the effects of medication. Some of the most common include:

- Bacteria that have cell walls may evolve into organisms that do not have cell walls, thereby evading the effects of medication that interferes with cell wall synthesis.

  Bacteria might alter their metabolic mechanism so that antibiotics could not sabotage it.

  Bacteria could also produce an enzyme that would neutralize antibiotics such as penicillin. They would break the lactam ring with an enzyme called penicillinase.

  Bacteria could expel an antibiotic through reverse pinocytosis before it took effect.

Fungal and viral organisms can become resistant through some of the same processes.

In summary, if you want to identify an infectious organism, first ask yourself where you found it and what were the symptoms. What part of the body? Did they have a fever, discharge, etc.? Then examine it visually (when possible), and compare it with known samples. E.g., suppose someone has a sore throat, and you suspect strep. You would visualize the organism and compare it with known streptococcus. If you are fairly certain that it is strep, perform a biochemical test by mixing with streptococcal antibodies to see if there is an antibody-antigen interreaction. Antibodies are very specific.

The same basic steps could be taken with viruses and fungal organisms.

Before we leave the topic of microbiology, there is one subject that should be expanded upon. What prevents the earth from being overrun by bacteria, fungi, viruses, and parasites? Sunlight, cold weather, and saltwater. All are effective at killing them. Sunlight is essential for photosynthesis and for killing dangerous microbes. Direct exposure of surgical instruments to the sun for ten minutes is considered an effective way of sterilizing them (as good as an autoclave).

If you dump raw sewage into ocean water, the bacteria will be dead by the time they get as little as two feet from the dump site. Cold weather is also effective.

Botany

Let us discuss a new topic:

Botany

Dr. William Withering, MD, FRS, FLS (1741–1799)

Jan Baptist van Helmont

George Washington Carver

Phototropism is the tendency of plants to grow towards sun light, such as a tree leaning towards the sun.

Geotropism is the tendency of plants to grow towards the center of gravity such as a hanging vine.

Hydrotropism is the tendency of plants to lean towards water such as a tree leaning towards a lake.

Botany is the study of plants. There are 6 million species of life on Earth and 391,000 of them are plants. As we said, the difference between plants and animals is locomotion.

There are two names that deserve recognition in the field of botany: William Withering and George Washington Carver. William Withering was a British geologist, botanist, and physician who lived in the eighteenth century. He had patients who suffered congestive heart failure (known as dropsy), who were largely untreatable. He noticed, however, that some were getting better. They told him that they were visiting a "witch" and that she was providing a potion. He visited the witch and gave her a gift in exchange for her potion. It turned out to be tea made from the flowers of foxgloves. The active ingredient, digitalis, did not become available in pure form until 1930. But even a crude preparation could slow the heart and increase its inotropy (strength of muscular contractions), leading to significant improvement. Today 30 percent of our medications still come from plants.

George Washington Carver came along a century later and made a significant contribution to the study of edible plants, especially the peanut. He also emphasized the importance of crop rotation and the use of legumes.

Most plants depend on soil for their supply of water and trace elements but make the bulk of their biomass from air, water, and sunlight. This was discovered by yet another botanist named Jan Baptist van Helmont, who lived in the 1600s. He took a barrel of soil and weighed it carefully. Then he planted a seed and nurtured it until it grew into a tree that weighed several hundred pounds. Then he carefully recovered the soil and reweighed it. Its weight had only changed by a few ounces.

He concluded that plants make the bulk of their biomass from air, water, and sunlight and take very little from the soil. So if you are growing an orchard you only need to place a few ounces of minerals on the ground.

There are four ways for plants to obtain nutrition:

- Some are saprophytes, which means "feeding on the dead." It refers to plants that break down preexisting food.

Others are parasites such as mistletoe. They live off other life forms, typically trees.

Some are photosynthetic, which means "making food and oxygen with light." This includes most green plants and provides living things with food. They are at the bottom of the food pyramid, a theoretical pyramid that indicates how all living things eat. Where does their food come from?

Some plants are carnivorous.

Photon of light: Let us talk about photosynthesis:

$$H_2O \longrightarrow H_2 + \tfrac{1}{2}O_2;$$

ultraviolet light.

The O2 goes off as atmospheric oxygen, and the H2 is used to convert NADH (nicotinamide adenine dinucleotide) into $NADPH_2$ (nicotinamide adenine dinucleotide phosphate in the reduced form):

$$NADP + H_2 + \tfrac{1}{2}O_2 \quad \rightarrow \quad NADPH_2 + \tfrac{1}{2}O_2 \text{ (atmospheric oxygen)}$$

This reduced molecule is then used to make sugar, a source of hydrogen (and energy) for all living things.

Carnivorous Plants

Some 750 species of plants are carnivorous and supplement their nutrition by catching insects. We will expand upon this later.

Let us think about plants in terms of how they

Reproduce.

Adapt to new environments.

Draw water into their leaves.

Store the food made with photosynthesis.

Protect themselves.

Survive cold weather.

Differ from animals: in what ways are their cells different?

Why are some plants carnivorous (catch insects)?

Which plants generate interesting substances such as LSD, mescaline, psilocybin, amphetamines, opiates, coumadin, belladonna, adriamycin (anticancer drug), vincristine (anticancer drug), vinblastine (anticancer drug), or barbiturates?

Why do some plants have broad leaves?

Why don't all plants reproduce? Examples are annual flowers, olive trees that do not produce olives, coconut trees that do not produce coconuts, date trees that do not produce dates, and crops that can only be harvested once and whose seeds will not germinate, as well as fruits or grapes without stones or seeds.

Why are some trees tall like the redwoods of Northern California, which are 300 feet tall and in the sequoia family?

How do plants reproduce? Let's make a list:

Some of the oldest life forms on Earth, such as mushrooms produce spores. They have a short life span, and when they die, their heads break open and release a fine powder that has the consistency of flour. Each grain is capable of germinating into a new mushroom. Spores have five advantages:

- They are very impervious to heat, cold, dehydration, and acids.

    They come in large numbers, assuring that the species will be perpetuated.

    Wind spreads them over a wide area, ensuring their continued survival.

    They can remain dormant for long periods, allowing them to escape unfavorable conditions.

They are also resistant to many herbicides, and for this reason it is difficult to get rid of them.

Other plants include vines, succulents, tubers, flowering plants, grasses, trees, and cacti.

Vines such as ivy spread along surfaces and establish an extensive root system. But they come from one central plant. Tubers have hollow stalks and will spread over ground, underground, and in water. And they too come from one plant.

These plants reproduce asexually. But plants that produce flowers reproduce sexually. The flowers' bright color and fragrant scent attract insects (especially bees) that help it to reproduce. Bright colors are caused by pigments called carotenoids and anthocyanins that contain double bonds. They resonate at specific wavelengths, thereby generating light with specific colors. The scent comes from esters which are formed when carboxylic acids combine with alcohols and are used in perfumes.

Insects carry pollen from the stamen of one flower to the pistil of another. Then the flower undergoes metamorphosis and becomes a fruit. If the fruit grows on a vine or a bush, it is commonly referred to as a berry, and if it grows on a tree it is a fruit. Most fruits contain seeds, and when an animal eats them, it encases them in excrement and drops them some distance from the original plant. This serves

- To protect the seeds against organisms that might destroy them.

  To place the seeds in a warm, moist jacket of fertilizer (to get them going).

To move the seeds away from the original plant so that they do not compete for root space and are disseminated over a large geographic area. This increases the chances of the species surviving.

Watermelons, gourds and pumpkins are berries with hard endocarps (shells).

Some plants will produce vegetables instead of fruit. The difference is that fruit contains seeds and vegetables do not. This could also be a means of reproduction and vegetables can sprout into a new plant.

Some plants have adapted to retain water; these include succulents and, more specifically, cactus. Succulents are like celery and rhubarb. They have fleshy tissues that store water. Cacti have thick integuments (skins), thin leaves in the way of needles, and a barrel-like body that allows them to store water for long periods. They also have an extensive mat of roots near the surface that can grab every drop of water quickly, in the event of a rainstorm.

Trees reproduce. But the hallmark of a tree is its long life. Some will live to be 10,000 to 12,000 years old. Some seagrasses that grow in tide pools are believed to be 100,000 years old.

Grasses such as hay, wheat, corn, rice, and bamboo (the largest member of the grass family) provide the earth with much of her food. They grow quickly and in a wide variety of soil and climate conditions.

Plants are on the lowest rung of the food pyramid and produce the food that most living things consume, either directly or indirectly. Green plants produce food through photosynthesis, which is carried out by organelles called chloroplasts. Chloroplasts will turn water in a flask green and will make oxygen while suspended. This was discovered by a British botanist named Joseph Priestley in the 1700s.

Food Pyramid

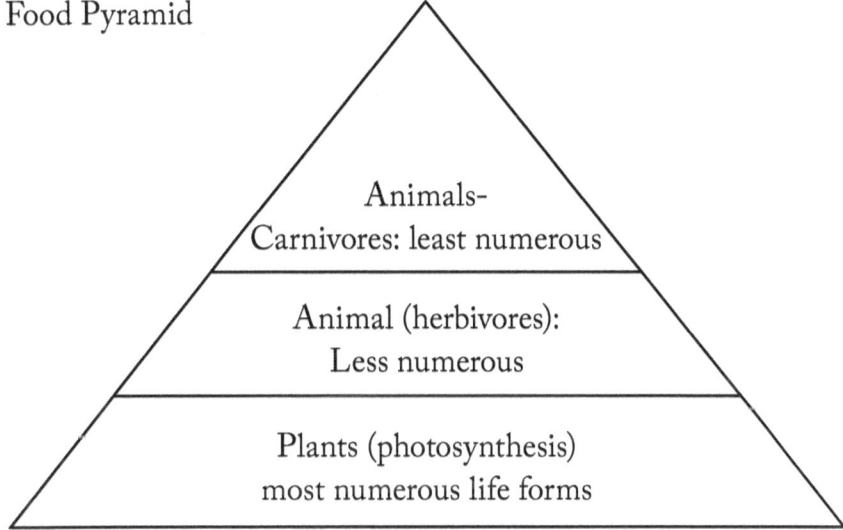

Food Pyramid

What makes seaweed float to the surface? Nodules filled with carbon monoxide. They act as balloons carrying the plant upward.

What element does seaweed contain that would make it dangerous to eat without proper preparation? Iodine.

Why is it difficult to be a pure vegetarian? Because plants are made of so-called low-quality proteins, which do not contain the same amino acids in the same proportion as animal meat. So over a prolonged period your body could become deficient.

Why do some plants produce toxins or substances that could be used as medication? To defend themselves against biological pests. One interesting example is the caffeine produced by coffee, tea, and chocolate. It paralyzes insects.

Why are some plants carnivorous? There are 750 known species of carnivorous plants. Most live in swamps and grow in shallow waters that are warm and mildly acidic (like a hot cup of tea) and the water is typically in motion. This has the effect of leaching minerals out of the soil. So the plant supplements its nutrition with insects.

And which plants produce interesting substances?

- LSD comes from the seeds of the morning glory flower and from the Ergotamine mold that grows on rye.

  Mescaline comes from the "buttons" of cacti that grow in the American Southwest.

  Psilocybin comes from a mushroom that grows in the Amazon.

  Cocaine comes from the coca plant of South America.

Opiates come from the opium poppy of Asia and the Middle East. They are a distant relative of ice poppies that also produce small amounts of opium.

Amphetamines come from the ephedrine plant in Asia.

Oxalic acid (used to make barbiturates) comes from the leaves of the rhubarb plant.

Adriamycin (an anticancer drug) comes from a seaweed that grows in the Adriatic Sea.

Vincristine (an anticancer drug) comes from a small white flower that produces vinca alkaloids.

Vinblastine (an anticancer drug) comes from a small white flower as above.

Strychnine comes from hemlock of the carrot family.

Coumadin comes from hemlock.

Belladonna comes from nightshade (related to tomatoes)

Digoxin comes from the flowers of foxgloves.

- Capsaicin cream comes from hot chili peppers, red and orange.

- Aspirin comes from the sap of the evergreen tree.

- Potassium cyanide comes from red beets.

So let's answer four final questions.

First, what is the secret behind plants that will not reproduce such as annual flowers, fruits without seeds or stones, coconut trees that do not produce coconuts, date trees that do not produce dates, cherry trees that do not produce cherries, olive trees that will not produce olives, and crops that will mature once and whose seed cannot be used to grow a new crop?

To understand this, we have to revisit genetics and the classification of living organisms. For an organism to reproduce, it must have an even number of chromosomes, e.g., 46. So when meiosis occurs each gamete receives the same number—in this case, 23. All members of the same species will have the same number of genes. But members of the same genus will have a different number of genes. For example, all horses will have the same number of genes. All donkeys will have the same number, but horses have a different number than donkeys. So let's say that horses have 46 chromosomes and donkeys have 44. A donkey and a horse can produce a mule. But the mule will have 45 chromosomes. He cannot produce one gamete with 22 chromosomes and one with 23. Both gametes have to have the same number, or they cannot form. So the mule will be sterile.

The same would be true of a tiger and a lion. They are members of the same genus. So they can produce a liger. But ligers are sterile. Any living organism with an odd number of chromosomes will be sterile. So the secret to creating organisms that cannot reproduce (including plants) would be to breed members of the same genus (but not the same species). If your orange tree and your lemon tree were pollinated by bees, you would get a fruit that was a cross between a lemon and an orange. It will not be able to reproduce and will not produce seeds. The same would be true for any other plant.

Certain sprays will prevent plants from reproducing; i.e., they can stop olive trees from making olives. But they work on a different principle. They are long-acting insect repellents that prevent bees from pollinating the plant.

So let's move on to our second question. Why do some plants produce a high concentration of certain desirable substances such as sugar, caffeine, or nicotine? To understand this, we have to revisit our unit on evolution and genetics. Remember that the purpose of a berry or a fruit is to be eaten so that the seeds can be spread to another geographic area. The more sugar the fruit (or berry) has, the more likely it is to be eaten. And remember that nicotine and caffeine are natural insecticides. They prevent the plant from being eaten. Remember too that evolution can take place quickly, as was true of Darwin's finches. So a plant that grows in an area where there are many insects might develop the ability to quickly create high concentrations of caffeine or nicotine.

But we can also use genetics. The concentrations of sugar, caffeine, and nicotine are influenced by additive genes (similar to the ones that affect stature). The more you have, the more you produce. So a plant with many caffeine genes, or many nicotine genes, will produce a high concentration of each. Finally we can use genetic engineering, which could help us in three ways. First, remember that plants tend to have a relatively small number of large chromosomes that can be manipulated under a microscope. It is possible to remove substance-forming genes from one plant and place them into another. So the seeds of a new plant could be endowed with several caffeine or nicotine-producing genes. Second, it may be possible to "turn on" a promoter gene. This would "promote" the activity of genes that create caffeine and nicotine. And finally (third), we might be able to suppress a "suppressor gene." We could prevent this gene from suppressing the genes that make caffeine and nicotine.

In summary, we breed plants until we have hybrids that naturally produce large quantities of caffeine or nicotine. Then we add more genes to its nucleus. And finally we might add more promoter genes and suppress the suppressor genes.

So let's address our third question. Why do some plants have very broad leaves? To answer this, we have to go back to our unit on ecology. Rainforests (like the Amazon) have a canopy of trees that does not allow much sunlight to reach the ground. Plants that live on the forest floor have to capture as much sunlight as possible to carry out photosynthesis. Since they have relatively little available they need a leaf with a broad surface area so as to capture as much light as possible.

Why are some trees tall, such as the redwoods of Northern California (in the sequoia family), which grow to 300 feet?

There are two answers:

1.
2.
3. Extremely tall trees are also old, on the order of 3,500 years, and they have a broad base of at least 20 feet. In the course of their lives they have been exposed to many fires. It is unlikely that even a hot fire would destroy the capillaries at the center of the trunk so there would be a survival advantage to being tall.

1. All green plants (including trees) carry out photosynthesis. The leaves of a tall plant would always have access to sunlight.

2.

3. We measure the total amount of light collected in terms of lumens. If a plant lives on the floor of a rainforest it will receive very few lumens per square inch and its leaves will need many square inches to receive enough light to run photosynthesis.

Robin Hill

Discovered that the oxygen from photosynthesis comes from water, not carbon dioxide (1937)

The Circulatory System of Plants

To understand the circulatory system of plants we need to know that molecules of water attract one another.

(H$_2$O)

(H$_2$O)

(H$_2$O)

Let us conclude by briefly discussing the plant's circulatory system.

Plants contain capillaries: very thin tubes that allow water and nutrients to circulate throughout their tissues. They work on the principle of surface tension and convection. Molecules of water adhere to one another, and as the water in the leaves evaporates it pulls more molecules of water behind it (see diagram).

There are two kinds of capillaries: phloem and xylem. Xylem brings water and nutrients, from the soil to all parts of the plant. Phloem brings substances synthesized in the leaves to the fruits of the plant (e.g., sugar). It typically takes one ton of water to create one pound of sugar. So sugar plantations and fruit farms need copious amounts of water.

Let us conclude by asking random questions:

Do plants breathe? Yes. They have pores on the surface of their leaves that expel oxygen during the day and bring it back by night. These are known as the "light and dark" reactions of photosynthesis, respectively.

Do plants have a way to protect themselves against dehydration, disease, and insect pests? Yes. The leaves and fruits are covered with a wax called cerumen.

Since plants do not generate heat, how do they survive cold weather? They contain unsaturated lipids that have a low freezing point. They remain liquid at low temperatures.

To say that a compound is "unsaturated" means that it contains double bonds and can combine with hydrogen. Gram for gram, they have less energy than "saturated lipids," which have single bonds and cannot combine with hydrogen. But they remain liquid even at low temperatures.

Most animal fats contain saturated lipids. But the animals of the Arctic have unsaturated lipids.

Let us move onto entomology. Entomology is the study of insects. Insects are exoskeleton life forms that do not have a spine, lungs, or a cardiovascular system. Their hearts are a single tube that pulsates, and they do not have arteries or veins. Their bodies are covered with chitin, a hard glycoprotein. It contains a series of holes that allow air to move in and out and fluids circulate as a result of their movements.

They eat almost everything and can use either ATP or ADP as a source of energy. They can see ultraviolet light, and those that fly defy laws of aerodynamics. Their body size seems too big for their wingspan.

The first life forms to leave the oceans were the scorpions, and finally there are spiders. Any exoskeletal life form with two body segments and eight legs is a spider by definition. All spiders are venomous. But most have fangs that are too short to penetrate human skin.

Some of the truly dangerous spiders include the brown recluse, who lives east of the Rockies, and the banana spider, who lives in the Caribbean. Brown recluse spiders inject a self-renewing venom that creates an ulcer that keeps getting bigger until it is surgically debrided. The banana spider is so poisonous that it can kill a cow. They are large brown spiders that live in banana trees.

The most dangerous spider in California is the black widow, which kills 4 percent of the people it bites, mostly infants. Its bite is painful and leaves the victim feeling as if they have a flu. There is an antidote, but it has side effects and should be avoided. The usual treatment includes a tetanus booster and an antihistamine such as Benadryl.

For its size, the black widow is the most poisonous creature on Earth. It is black and has a red hourglass on its belly. It is the female who is aggressive and poisonous.

Let us consider another insect: bees. Thirty percent of our crops depend upon them, and their numbers are diminishing. As atmospheric temperatures rise, the number of bees is going down.

If you are attacked by bees (which is rare), you have to receive at least 300 stings before the venom threatens to shut down respiration (unless you are allergic to them). If you are stung, try to remove the stinger without squeezing the sac at the end of it. You can avoid injecting yourself with the venom, and if you have an allergic reaction, carry an EpiPen and inject yourself with 0.3 cc of epinephrine. If you are over sixty, be careful that this does not trigger atrial fibrillation. Your doctor should do an EKG before prescribing it.

The venom of Africanized bees is no more deadly than that of any other. It is just that they are more aggressive and are more likely to attack in swarm.

Another group of "social insects" (working in groups) are the ants. They keep our world clean and help to eliminate dead matter. They dig intricate tunnels that might be as long as two miles and send scouts in search of food. If it is found, a message is sent to the colony. So if you do not want an ant problem, be sure that the scouts do not find food.

And finally there are insects that devour whole crops like locusts and cicadas. Fortunately they do not come often.

So insects pollinate plants, eliminate carrion, and reduce the number of harmful organisms such as flies. And for their size, they are the fastest organisms on Earth. It is estimated that if most insects were the size of humans they could move at 260 mph!

Nutrition

Let us shift our attention to nutrition. We consume seven basic foods:

- Carbohydrates

  Proteins
  Lipids (fats, oils, cholesterol)
  Vitamins
  Salt
  Roughage (foods with fiber such as lettuce and celery)
  Minerals (e.g., iron, magnesium, and zinc)

Carbohydrates provide energy. They produce a unit of energy called a "calorie." A calorie is simply a unit of heat. A small calory will heat 1 cc of water by 1 degree Centigrade. A large calory will heat 1,000 cc of water (one liter) by 1 degree Centigrade. The calories of foods are reported as large calories.

A small calorie is also 4.18 joules of energy: the energy required to accelerate 4.18 kg. (about 9 pounds) at a rate of one meter/second^2 2 (meter per second per second) over a distance of one meter. If you think about this it means that it is pretty hard to burn calories by exercising. It takes one calorie just to accelerate 9 pounds at a rate of one yard per second squared over a distance of one yard. Nevertheless, it is easier to lose weight if you engage in aerobic exercise for twenty minutes a day. You need to speed your heart up by 20 percent for twenty minutes.

Scientists determine the calorie value of foods by placing them into a device known as a "bomb." It makes the food combine with oxygen and measures the heat that is generated. It is a process known as food calorimetry.

If we do not consume carbohydrates, we convert protein into carbohydrates through gluconeogenesis. This takes place in the liver and means "forming new glucose" in Latin. If we have too much carbohydrate, the excess can be converted to lipids and stored in adipose cells. A person who is of normal weight will have approximately 200 million adipose cells. A person who is obese will have more than a billion, and once formed, these cells are hard to get rid of.

So protein can be converted into carbohydrates, and carbohydrates can become lipids. But these processes cannot run in reverse. And there is one more fact that we should consider. If you metabolize lipids without carbohydrates, you will develop ketoacidosis. So carbohydrates play a key role.

In general proteins provide the building blocks for new proteins, carbohydrates provide energy, and lipids serve as insulation. But the roles are interchangeable.

So what are vitamins? There are four ways to answer this. To a biochemist they are the precursors of coenzymes: substances that enzymes need to function. You could think of enzymes as a jigsaw puzzle that needs a missing piece. These missing pieces are coenzymes.

Another way is to think about the actual role of each vitamin.

- Vitamin A is a radical sink that neutralizes radicals.

  Vitamin B is essential for the CNS.

- Vitamin C is a reducing agent that repairs connective tissue. We develop scurvy if we do not have it.

- Vitamin D transports calcium to bones and teeth. The incidence of muscular dystrophy is also higher in patients who are deficient in vitamin D for reasons that are not clear. We will develop rickets (bow legs) if we do not get enough.

- Vitamin E is also a radical sink.

  Vitamin K is involved in blood clotting.

And we should think about vitamins in terms of whether they are water-soluble or fat-soluble.

Vitamins B and C are water-soluble, and Vitamins A, D, E, and K are fat-soluble. This is important because vitamins can be toxic if we overdose on fat-soluble vitamins. If they are water-soluble, the body will remove them by dissolving them in the urine. But fat-soluble vitamins can accumulate.

And a fourth way to think of vitamins is to identify their source:

> Vitamin A: green leafy vegetables
> Vitamin B: meat, fish and poultry products
> Vitamin C: citrus fruits, cranberries, and strawberries
> Vitamin D: synthesized by our skin when we receive sunlight
> Vitamin E: green leafy vegetables and eggs
> Vitamin K: synthesized by bacteria in our gut

We also need salt (NaCl). It does four things:

1. Sodium ions and chloride ions are involved in neurotransmission. As neurological impulses travel along a nerve fiber, they trigger a wave of sodium ion depolarization followed by a wave of potassium ion repolarization. When a pore opens on a nerve fiber, it allows sodium ions to enter. This changes the internal voltage from negative 90 millivolts to positive 30 millivolts. That causes another pore to open, and the process repeats itself. The wave travels down the nerve at 1,200 feet per second.

   In the brain there are chloride channels. When they allow chloride ions to enter, they increase the negative charge inside the nerve fiber, thereby stabilizing the neuron and making it more difficult to fire. This creates a sense of euphoria, tranquility, and sleepiness. This is caused by a neurotransmitter called GABA (gamma amino butyric acid), whose concentration can be increased by benzodiazepines such as Valium or Xanax.

2. We need salt to perspire and to prevent the mucus in our bronchial passages from becoming too thick. Perspiration is essential for maintaining correct body temperature (98.6° F), and mixing water with mucus is essential for us to be able to breathe. Without it we would develop cystic fibrosis. Without treatment patients, with cystic fibrosis rarely live beyond thirty-five.

   Sweat glands in the skin and mucous glands in the bronchial passages secrete salt first, and water follows. Water is drawn to the surface of these glands by osmosis.

3. We need chloride ions to make the hydrochloric acid in our stomachs, which is essential for the absorption of nutrients, killing dangerous organisms and reducing the risk of stomach cancer.

4. And finally, salt helps us to retain water. It prevents us from either perspiring or urinating too much. It holds water inside our bodies through osmosis.

Roughage (foods that contain fiber) reduce the risk of colon cancer by allowing stool to pass through the GI tract with greater ease.

Minerals (such as calcium, potassium, sodium, iron, magnesium and zinc) play multiple roles.

1. Calcium is involved in:

   A) bone and tooth formation
   B) clotting of blood
   C) muscular contraction
   D) Can sometimes prevent aneurysms from expanding

2. Sodium and potassium are involved in neurotransmission.

3. Iron plays a role in the manufacture of red blood cells. It is also part of cellular respiration.

4. ATP cannot function without magnesium. The kidneys also need magnesium to retain potassium, and muscles need it to relax. It also plays a role in the function of the CNS. If patients are deficient, they will develop cognitive deficits. This is common with alcoholics.

5. Many trace elements, such as zinc, act as cofactors for enzymes. They provide a missing piece, much like coenzymes, and the enzyme cannot function without them.

We should think about nutrition in two other ways:

1. How many small calories do we need?
2. How can you be sure you are consuming a balanced diet?

First, we have to decide how much we are supposed to weigh. There are two ways to do it.

1. We could use the New York Life Insurance system. You determine ideal weight by measuring the patient's height in inches. And you say that if they are an adult male they should weigh 110 pounds for the first five feet and 6 pounds for every inch after that. If they are an adult female they should weigh 100 pounds for the first five feet and five pounds for every inch after that.

   If they are five feet and 10 inches (and male), their ideal weight would be 170 pounds; if they are female, 150 pounds. But there are two additional factors:

   A) You have to decide if the patient has a small frame, an intermediate frame, or a large frame. It is based on the circumference of the wrist. If they have a slender wrist, they have a small frame, and you would deduct 10 percent from ideal weight. So for the small-framed patient above, the ideal weight would be 153 pounds if they were a male, and 135 pounds if they were a female. If they had a medium frame, you would keep the 170 pounds if they were male, 150 pounds if they were female. If the wrist diameter and frame were large you would add 10 percent and the ideal would be 187 pounds for a male and 165 pounds for a female.

B) So let's say that the ideal weight is 187 pounds. Now we create a range in which we give or take 10 percent and we say that the ideal weight is in the range of 170–204 pounds for a male and 150–180 pounds for a female. Anything within this range is okay.

2. The second (and perhaps more accurate) method would be to calculate the BMI (body mass index). Again we have to know the patient's gender and height. Then we use the BMI tables (which can be found on the Internet) to calculate their ideal weight. An emaciated person would have a BMI of less than 18.5. Ideal would be a BMI of 18.5-25. An overweight person would have a BMI of 25–30. An obese person would have a BMI of 30–35, and a morbidly obese person would have a BMI of more than 35. This is the system that is most frequently used today.

So on the basis of the above, we could calculate ideal weight. The number of small calories consumed per day divided by 10 will equal the weight that they will ultimately achieve (and maintain) if they continue to eat the same amount. So let's say that they consume 1,800 calories a day. They will continue to lose weight until they weigh 180 pounds, and they will stay there. They would lose one pound for every 3,500 calorie deficit. So let's say that they normally weigh 250 pounds and consume 2,500 calories a day. If they dropped down to 1,800 calories it would be a deficit of 700 calories and they would lose one pound every five days: 3,500 calories/700 calories/day = one pound every five days. Over a prolonged period the patient would not want to lose more than two pounds a week. It would be too much of a stress on the body.

We need a minimum of 600 calories a day to run our essential functions, such as breathing, pumping blood, running our kidneys, etc. Anything beyond that would be for our personal needs.

So there is one more topic. How can we be sure we are getting a balanced diet? There are two ways:

1. The best way is to eat:

    A) One serving of meat, poultry or fish each day
    B) One serving of a grain product each day
    C) One serving of a dairy product each day
    D) One serving of fruit each day
    E) One serving of a green leafy vegetable each day

If you eat these foods you would not need a vitamin pill.

2. Most recently nutritionists have stressed diets rich in:

    A) Fish
    B) Whole grains
    C) Fruits

It is believed that this provides the best chance of prolonged survival without Alzheimer's, Parkinson's, or cancer.

Miscellaneous facts

Let us talk about miscellaneous facts in biology.

Plants that live in rainforests receive limited sunlight. The floors of rainforests are shaded by the canopies of trees and are dark. So the number of lumens received per square inch is small. Therefore the leaves are going to need many square inches to collect sufficient light. Hence a large leaf.

Earlier we spoke of pasteurization and vaccines. Let us explore each in more detail. Pasteurization is the process of killing dangerous microbes within food and beverages through the application of heat. It can be done in two ways:

We could heat foods and beverages, to near the boiling point of water (100 degrees Centigrade or 212 degrees Fahrenheit) for ten minutes. Or we could expose them to the "flash" temperature (300 degrees Fahrenheit or 149 degrees Centigrade) for 15 seconds. Alternatively you could expose the product to gamma radiation. This is usually effective, although some viruses can tolerate more than 200,000 rads.

How do vaccines work? They present the body's immune system with the antigens (surface proteins) of dangerous pathogens so that it can develop immunity, which typically requires ten days. Vaccines have been around since the days of the ancient Chinese, when physicians would take scabs from patients who had recovered and place them into the noses of people who needed protection. But there are four things that we should know about vaccines:

- Many have to be maintained at a specific temperature, typically 35–45 degrees Centigrade. Too cold or too hot can destroy them.

  They have an expiration date.

  About once every ten years the patient should receive a booster, a conventional vaccine that has been diluted ten times.

Some vaccines have to be repeated every few years because the organism changes. E.g., flu shots should be given annually, and pneumonia vaccines should be repeated every five years.

Vaccines come in two basic forms: "live" vaccines (consisting of an organism that is alive in a weakened state) or "dead" vaccines, in which the organism is dead or the vaccine only uses its surface antigens.

As a rule, it is okay to give "dead" vaccines to immunocompromised patients. But they should not receive "live" vaccines. These might trigger an active infection in a patient with impaired immunity. If an immunocompromised patient needs a "live" vaccine, they should receive it before they become compromised, e.g., prior to having a splenectomy or starting immunosuppressive medication.

How do we measure blood pressure? We need two instruments:

- Sphygmomanometer
- Stethoscope

A sphygmomanometer is an inflatable cuff that we pump with air. A stethoscope is a listening device with earpieces and a bell, which doctors and nurses use to listen to hearts and lungs.

The investigator places the sphygmomanometer on the upper left arm and the stethoscope in the median cubital area (opposite side of the elbow). Then they inflate the cuff until they can no longer hear a pulse. This tells them that the median artery has been temporarily occluded. The investigator deflates the cuff until a pulse becomes audible. The first sound will be at the systolic, the higher of the two numbers.

The sphygmomanometer has a meter that will tell them the pressure. Then the investigator continues to release the pressure and the sound disappears. As the pressure continues to drop the sound of a pulse will reappear. This is the diastolic (lower number) and is read off the meter.

Electric cuffs do the same thing with electrical components.

The Future

Let us conclude by considering some of the problems that are developing and what we must do in the future. There are almost eight billion people on Earth, and we depend on other life forms for food, oxygen, and 30 percent of our medication. We are endangering the environment in four ways:

1

Temperatures are rising as a result of greenhouse gases especially carbon dioxide. These gases absorb and retain the sun's heat. This is reducing the amount of oxygen in the oceans, affecting the health of all living organisms, and causing the polar ice caps to melt. Melting the ice caps will trigger a chain reaction. The ice reflects much of the sun's heat back into space. Without them we will get even hotter, and we will eventually become very warm. As the oceans warm up, they will lose their ability to dissolve carbon dioxide. It will stay in the atmosphere and make us even hotter. This will affect all life. Remember that enzymes lose function as temperatures change.

2

The pH of all water is dropping. Carbon dioxide is an acid anhydride and combines with water to form acid. Changes in pH will also have a profound effect on life. pH has an effect on enzymes and upon solubility.

3

We must be careful with the substances that we put into the environment. Most plastics will not degrade for hundreds and possibly thousands of years. Other wastes such as iron and phosphates could trigger a devastating overgrowth of algae.

4

As populations grow we are destroying the natural habitats of other life forms.

All of these factors could devastate life that we depend upon.

Conclusion

So now we both know a few secrets of life. I hope you enjoyed this as much as we did.

Nicholas J. Orme, MD
Atif Elnaggar, PhD

Index

A

abdomen, 224, 301, 311
acetylcholine, 40, 226
achlorhydria, 261
acid anhydrides, 160–61, 374, 445
acid-base theory, 144, 147
acidosis, 78
acids, 145, 161
  acetic, 146, 161
  carbonic, 146
  carboxylic, 419
  hydrochloric, 145, 261
  nitric, 145–46
  strong, 145
  sulfuric, 145–46
  weak, 145–46
adenosine triphosphate (ATP), 18, 73,
  432, 438
adenovirus, 391
adrenal glands, 220
African bees. *See* bees
algae, 403
  blue-green, 4, 403
alleles, 170–71, 174
alveoli, 41, 248–49, 257
amide bond, 61, 63, 65, 81, 86, 95, 104,
  106, 118, 154
amine group, 114, 117
amino acid, 2, 61, 63, 65, 71, 86, 114,
  117–18, 423
  C-terminal, 61, 63, 65, 85–86, 104,
    114, 117–18
  N-terminal, 61, 63, 65, 85–86, 104,
    114, 117–18
amoebas, 6, 50, 403

amphetamines, 417, 424
amphibians, 200–201
Anaphase, 174, 187, 190
anatomy, 215, 217
anion, 110, 112, 131, 145, 147, 155
Antarctic Circle, 336–37, 340. *See also*
  South Pole
anthocyanins, 419
antibiotics, 102, 296, 330, 397, 405,
  407–8
antibodies, 320, 324–25, 327, 392, 408
  IgG, 324
  IgM, 324
antifungals, 257, 407
antigens, 285, 324–27, 392, 442–43
aorta, 220–21, 223–24, 260
aortic arch, 218, 220, 223
aortic valve, 218, 223, 238
Arctic Circle, 336–38, 340, 431. *See*
  also North Pole
arctic fox, 338
Arctic Ocean, 364
arms, anatomy of, 306, 310
Arrhenius, Svante, 144
Arrhenius acids, 144, 148
arteries, 13, 218, 220–21, 223–26, 249,
  261, 268, 281, 295, 309, 311–12,
  431, 443
  brachial, 223, 310–11
  gastric, 261
  iliac, 224
  lingual, 260
  pulmonary, 218, 223
  spinal, 224
asthma, 250–51, 258
atom, 110, 127, 135

cranial nerves, 226, 259, 273, 288, 290, 294, 309

Crick, Francis, 52, 55

crustaceans, 200, 363

cyanide, 102

cysticercus, 296

cystic fibrosis, 437

D

Darwin, Charles, 193–95
Origin of Species, 193

daughter cells, 162, 169–71

da Vinci, Leonardo, 216

dendrites, 40

dentate line, 267–68, 271

dermatomes, 226

dermis, 335

Devonian period, 193, 196

diabetics, 235

diet, balanced, 439, 441

diffusion, 142–43

digestion, 35–36, 260–61

digestive systems, 259

dipeptide, 104

diploid. *See* mature cells

diseases, 390
infectious, 297, 398, 401
obstructive lung, 254
restrictive pulmonary, 251

DNA, 13, 52, 64, 122
genetic code, 66, 71
structure of, 52

dogs, 210

dropsy. *See* congestive heart failure

drugs
anticancer, 328, 330
antiviral, 393, 407
fertility, 170

E

earth, 2, 193, 336, 415

Earth, atmosphere of, 1–2

ecology, 336

ecosystems, 336
chaparrals, 349
deserts, 360
marshes, 363
rainforests, 427, 441
salt flats, 336, 363–64
tide pool, 363

ectoderm, 13

EKG, 232, 245

electricity, 239

electrolytes, 112–13, 126

electrons, 78–79, 110–11, 124, 130, 135, 151, 392
unbalanced, 134

Electron Transport System (ETS), 73, 78

elk, 341

Embden-Meyerhof-Parnas pathway, 73, 78

emphysema, 250, 257

endocrine systems, 313

endoderm, 13

energy, 4, 18, 46, 54, 61, 63, 73–74, 87–88, 99, 104, 106, 114, 372, 431–32, 434–35

energy production, 73

entomology, 431

enzymes, 85, 101, 107, 119, 154
activity of, 85, 101, 152–53

enzyme-substrate interaction, 107

epidermis, 335

epiphyseal plates, 30

epithelium, 260, 271–72

esophagus, 260–61

ester bonds, 81, 94

esterification reaction, 81

parasympathetic, 226, 235, 268, 272, 294–95

sympathetic, 217, 226–28, 235, 272, 283

neurons, 13, 24, 40, 42, 285, 437

neutrons, 110

nicotine, 426

Niels Bohr model, 110, 130

nitrates, 145, 153, 239

nitrites, 70, 261

nitrogen, 1, 4, 61–62, 71, 87–89, 96, 104

noncompetitive inhibition, 101–2

normal respiratory effort, 250

North Pole, 337

nuclear membrane, 18

nucleolus, 17

nucleosomes, 71

nucleus, 110, 130

nutrition, 1, 98, 214, 323, 416–17, 423, 434, 439

O

oceans, 2, 4, 364–65, 374

octopus, 214

oncogenes, 192

order, 200

organelles, 99

organisms, ix, 4, 170, 177, 193, 197, 199, 207, 211, 272, 297, 403–4, 408, 425, 443

dangerous, 438

fungal, 408–9

Fungal, 408

infectious, 397, 408

multicellular, 196

salient, 404

single-celled, 73, 196

viral, 408

organogenesis, 17

organs, 17

Origin of Species (Darwin), 193

osmosis, 142–43

ostriches, 206

ova, 11, 170–71

ovaries, 169, 314

human, 170

oxygen, 4, 374, 417

P

pancytopenia, 327

parathyroid glands, 314

Parotid glands, 260

Pasteur, Louis, 375, 382

pasteurization, 442

penguins, 340

penicillin, 397, 407–8

pepsin, 261

peptide, 104, 123

Peyer's patches, 35–36

pH, 155, 157, 161

phenotype, 193

pheromones, 315

photosynthesis, 416

phylum, 198–99

physiology, 217

pia mata, 287

Pinocchio (Collodi), 1

pinocytosis, 47–48, 98, 315–16, 408

Pka, 155

Pkb, 155

plague. *See Yersinia pestis*

plant leaves, 427

plant reproduction, 425

plants, 416, 419–20, 423, 426, 441

carnivorous, 199, 417, 423

circulatory system of, 430–31

green, 416, 420, 428

oldest, 418

substances from, 423–25

plasma, 323

plasma cells, 325

www.ingramcontent.com/pod-product-compliance
Lightning Source LLC
Chambersburg PA
CBHW021347210526
45463CB00001B/9